Mom's Hecker Dress

Childhood Memories of Our Family and Farm Life

Kathy Murray and Judy Meyr

Printed in the United States of America

Contents

Preface

"Come on, Judy, let's get in the car. Mom's got her Hecker dress on; we must be going to town." After waiting a long time, Judy and John finally got out of the family car. They found out that the reason Mom had put on her Hecker dress was because Aunt Florence and Uncle Hans from Michigan had come to visit. Living on a farm in the mid 1940s and '50s meant you had work clothes and a little nicer outfit if you were going into town—either the little town of Hecker, Illinois, two miles away or the bigger town of Red Bud, five miles away. Mom had several house dresses she sewed using the same pattern. Yes, dresses to work in! Women in the '40s and '50s rarely wore pants, jeans, or shorts. Mom had a nicer dress that she reserved for going to town or for when company came to visit. That's the one John referred to as her "Hecker dress." She also had a "Sunday dress" for church. Dad, on the other hand, wore a suit for church, but the rest of the time he was in bib overalls.

This story of the Hecker dress reflects on life on our dairy farm in the 1940s and '50s. Our grandchildren sometimes ask us to tell them a story about when we were growing up. So, we decided to write this book so they can better appreciate both the past and the present.

Mom and Dad were both very thrifty and service minded. They were not demonstrative in showing affection, but their hearts were very full of love. As you read this book, you will see that times were different for both our parents and for us. We did not have all the modern conveniences and possessions that are taken for granted now. Neither were there the advances in medicine, so life expectancies were shorter. However, faith in God and love for family saw our parents through then, just as they do for us now.

Even though our sister Lois's life was cut short, she contributed in her own special way to this book. The water color drawing of the farm on the front cover was done by her. Judy designed the cover, and Kathy came up with the title. So, in a sense, this book is a labor of love from all three of us.

We would also like to acknowledge and thank Rita Eckert for her help in editing the book; John Hose for his help with the cover; and Sarah Hertwig, née Steiner, for her work in formatting. In addition, a big thank you goes out to all our family members for their contributions of stories and photos. We could not have done this without you!

With all our love,
Kathy and Judy

CHAPTER 1

Our Parents' Backgrounds

To provide a better understanding, we are starting way back at the beginning with some background information about our parents. . . .

Carl's Background

Carl Edward Guebert was born on a dairy farm on the outskirts of Red Bud, Illinois, on January 4, 1914, to Charles and Rosa Loesche Guebert, who were married in 1891. Charles's parents (Carl's grandparents), William and Engel Nagel Guebert, had both emigrated from Hanover, Germany, in 1850; landing in New Orleans, they went up the Mississippi River to St. Louis. William, his brother Henry, and Engel's father took a ferry across to Illinois, and since they had used all their money for traveling expenses, they actually walked the 40 miles to Red Bud! Friends in nearby Horse Prairie lent them a team and wagon so they could drive back to St. Louis to get the rest of their group and belongings.

In 1854, William's brother, Christian, traveled to America with their parents. Then in 1855, William and Christian together bought the 240-acre Henry McSimmons farm at the southeastern edge of Red Bud where William had been working for the last 5 years. The two brothers divided the farm in half, and William took the northern half.

In 1857, 27-year-old William and 16-year-old Engel Nagel were married. They prospered in finances and in numbers, having 13 children! They built a new house and a barn sometime in the 1860s. William and Engel also helped build a Lutheran church in Red Bud and were active in the church, as were their children. Carl's father, Charles, was the 6th child born to William and Engel on March 7, 1868.

When Charles (Carl's father) married Rosa, he took over the home place, making improvements as he went along and keeping the 120- (later 142) acre farm in good shape. They had 10 children, 8 boys and 2 girls. "Rosa was never very strong and with that flock of boys hers was indeed a busy life. Many a gallon of apple-butter had to be cooked and many a large loaf of bread had to be baked in that out-door bake-oven to keep those hungry mouths fed. Her name was Rose and she was a lover of roses. Some of the stocks she planted are still to be found in the garden" (quoted from *A Family Biography* by Arnold Guebert, written in 1936).

William and later Charles Guebert's home place where Carl was raised.

Wedding picture of Charles and Rosa Guebert, 1891

Carl Edward Guebert, our father, was the youngest of the 10 children. There was a difference of 22 years between Carl and his oldest sibling, Walter. We will start with Walter and work our way down the line of siblings to Carl. Most of the siblings had interesting nicknames and varied careers.

Walter, the oldest (1892–1962), had the nickname "Walks" and worked for the American Furnace Factory in Red Bud. He married Alma Ohlau and they had 13 children: Ralph, Lee, Anita,

Esther, Edith, Oliver, Alice, Helen, Clyde (Tom's sponsor), Rose (Kathy's sponsor), Jane, Carol, and Charles. (After Judy graduated from high school, she also went to work at the furnace factory. At noon, she would walk with Uncle Walks back to his house where Aunt Alma would have a big meal waiting for them.) Walks and Alma lived right next to the railroad in the northwest corner of town. Aunt Alma was kind of like a second mother to our mother Esther. She had long gray hair that she braided and then wrapped around her head like a pretty crown. She always had a quilt set up in a frame in her front room and would quilt whenever she had a little spare time. When we kids were small, Mom often dropped us off at Aunt Alma's while she went grocery shopping.

Next came Paul, born in 1894. He attended Walther College in St. Louis. He was working in the office of the Becker Milling Co. in Red Bud when he died of typhoid at the age the 20 in 1914, the same year Carl was born.

Arnold ("Dick") was born in 1896 (died in 1970), studied at Concordia College in Milwaukee and then Concordia Seminary in St. Louis. He married Anita Burgdorf in 1921 and they had 5 children: Paul, Robert, Lowell (Roger's sponsor), Clifford, and Marjorie. Uncle Dick was a Lutheran pastor and professor at Concordia College in Edmonton, Alberta, Canada, and was the author of the first Guebert biography written in 1936. Dick served as a mentor to Carl, writing him letters of advice and encouragement from Canada. The following letter is one Dick wrote Carl in 1939 when Carl was living on his own and had some reservations about what he had chosen to do.

Oct. 30, 1939

My dear Carl,

Your letter at hand. First the business. I'm taking for granted the Quarry Co. is being dissolved and that there is good reason for it. I am depending entirely on the judgment of you men down there, and have, therefore signed the receipt. When you send the money please send it by bank or postal money order as it is easier to collect here. Believe me, I sure appreciate this extra financial help at this time as I was already thinking of asking for an advance on my salary due to extra expenses the past month.

Now as to your great problem. I don't believe it is necessary for me to go into the matter and bring all kinds of arguments with regard to the farm and its location. You yourself say those are "probably only lame excuses" and the main trouble lies with you. That is also the conclusion I have reached after reading your letter. Now, Carl, in the first place let me say that you are taking the whole matter entirely too serious. You have magnified your troubles until they seem so large as to overwhelm you. That is a failing we all have when we are down in the dumps, but that wears off in time. Between me and you, Carl, I felt terribly homesick too when I first left home. Also when I first got to Spencer I felt the weight of responsibility and was want to wince under the load, but after a time you adjust yourself to the new surroundings and everything goes along swimmingly. And that's my advice to you at this time: Don't give up the ship during a spell of impatience. What you have taken on is not anymore than thousands of others have undertaken and many of them with less resources than you. Boy, how I'd like to be out on a farm, away from the turmoil of the city and close to nature. Of course I'd have enough reading material, and that's what you must have. Then invite friends in from time to time and you'll find the old place isn't so bad after all.

Think too of the thousands of young men in Europe today who are living in dugouts, who are chased from their homes and often beaten to within an inch of their lives. How wonderful that we can live in a free country where we and our loved ones are safe. – My advice therefore again: Stick it out for at least a year. You don't have to stay there all your life. After that if you still feel you don't like it you can make other arrangements. That will also give your land owner a chance to get another renter. Undoubtedly you have entered into certain obligations with him which you are in duty bound to meet.

The Lord is putting you through a trial, but don't forget: "He is faithful and will not suffer you to be tempted above that you are able." He is still the "Good Shepherd, who leads you beside the quiet waters, and whose rod and staff will comfort you, so that goodness and mercy shall follow you all the days of your life." Ps. 23.

Now another question: what about her? You haven't said anything in your letter about that very important angle. Don't be afraid to open up your heart to me. I assure you the confidence will not be misplaced. As far as your life is concerned much depends upon that matter. If you can look forward to the time you can lead the woman of your heart's choice to a home you are preparing for, it will make all the difference in the world. -- Now let me hear from you again. Don't hesitate to write again and as often as you feel the urge.

With sincerest love,
Dick

Since Uncle Dick and Aunt Anita lived in Canada, we did not know them very well. They came back to Red Bud in 1946 to celebrate their 25th wedding anniversary. Carl and Esther hosted the celebration at their newly bought farm. Following this paragraph are some pictures from that celebration. Even though we did not see much of Uncle Dick and Aunt Anita, we did get to know some of their children, as several of Uncle Dick's sons became ministers. They would stay with us when they came to St. Louis for classes and such at the seminary in St. Louis.

These are all the descendants of Charles and Rosa Guebert, Carl's parents, in addition to the Burgdorfs from Anita's side of the family.
Photo taken in front of original porch, south side.

Carl and his siblings at Dick and Anita's anniversary gathering in 1946.
Victor, Walter, Carl, Norma, Arnold, Alfred, Lorenz.
Photo taken on west side of farmhouse.

Aug. 1946 - Celebrating Dick & Anita's 25th

Anita Edith Alice John Joy Paul Oliver Lee
Cliff Vicki Evelyn Steve/Doris Rich Eileen Rosie Helen Clyde Jim/Don
Glen Lorraine Ann B. Carol Mary Ann Marj Grace Jane
Lois Barb B. Judy Suzanne Charlie

Carl's brother Alfred (Steve's sponsor) was born in 1898 (died in 1981) and was called "Bank" because he once said "unter de bank" for "under the bench." Alfred farmed and is the one who took over the Charles Guebert homestead. He married Norma Burgdorf (sister to Anita, Uncle Dick's wife) and they had 6 children: Doris, Eileen, Evelyn (Kathy's sponsor), Donald, Glenn, and James. With two Normas in the family, Bank's wife was called "Big Norma" and Carl's sister Norma was called "Little Norma." Big Norma was much larger than Little Norma, and we kids didn't think much about it. However, we don't know what Big Norma thought! Big Norma died in 1960, and the visitation was at the farmhouse. Brother Roger said he remembers Esther (our mother) crying at the funeral. It upset him so much to see his mother crying that he later prayed that she would never have to experience any of us kids dying. God answered his prayer. After being widowed for 8 years, Uncle Bank married widow Edna Kreyline in 1968.

Victor ("Vic") Guebert, born in 1900 (died in 1969), attended Concordia College in Milwaukee for two years and worked as a broker for Reinholdt & Gardner Investment Firm in St. Louis. He married Ruth Schuette (Lois's sponsor) in 1923, and they had 3 children: Victoria, Richard, and Suzanne. He passed away in 1969. The one thing we remember about Uncle Vic is the smell of the pipe he often smoked, especially when playing pinochle. Aunt Ruth was always busy and bubbly. They would sometimes host gatherings for the whole clan, and Aunt Ruth would bustle around the kitchen being sure everyone was well fed.

Clarence Guebert died as an infant the same year he was born in 1902.

Lorenz (John's sponsor) was born in 1903 (died in 1983) and acquired the nickname of "Honey" as a young boy because he was such a good boy as a child! He married Esther Rehmer, and they had 2 children: Mary Ann and Lorraine. Honey worked for the St. Louis Milk Co. and drove a milk truck, delivering glass bottles of milk to individual homes. Uncle Honey and Aunt Esther had a really neat old house with a wrap-around porch in Red Bud, and they had a huge roll-top desk. They lived right along the annual Red Bud 4th of July parade route, so that's where we often went on the 4th of July. Uncle Honey was great about including us kids in various excursions, realizing Dad's farming commitments. He took Judy and John to the old Highlands Amusement Park in St. Louis once. Judy remembers getting sick on the "barrel ride." The rest of us were too young for that trip. However, some of us younger ones got to go to Meramec Caverns with him. Dave and Tom once stayed overnight with him so they could go along as he delivered milk early the next morning. Later on, when Kathy was the only one left in grade school, she stayed overnight a couple times so that she could attend school basketball games.

Sister Johanna was born in 1906 and died of diphtheria when she was only 11, and Carl was 3 years old.

Sister Norma (Judy's sponsor) was born in 1908 (died in 1994) and was very close to Carl and to our mother Esther. For a year or two she worked in California. During that time, she wrote several letters to Carl addressing him as "My Dear Chunky" and signing them "Lovingly, Norma." Norma actually kept house for Carl for a while when he was farming on his own at the Kloepper farm before he was married. In Carl's birthday greeting to Norma on her 75th birthday, Carl wrote "Best regards to a faithful sister who stood by me when I needed her most back in 1939 out on the 'hill.' Although it was probably 'Blue-Bell' and that mile and a half ride every day to fetch the mail that attracted her more than I did."

Norma worked most of her life as a child care provider and companion for a Jewish family in St. Louis in the Park Plaza Hotel. She moved back to Red Bud when she retired. Norma never married but was a favorite aunt to her many nieces and nephews. All of us kids looked forward to visits from Aunt Norma. She would come and stay overnight and tell us stories of her travels all over the world. She loved taking cruises on the huge Queen Mary and Queen Elizabeth ships. Just making the beds, which we never did otherwise, would be fun when we did it with Aunt Norma! As a teenager, Judy spent a week with her in St. Louis when Aunt Norma had broken her arm. Judy helped her on the job working for the wealthy Scharf family. It was almost like a vacation, because Judy got to ride a street car and go to a Cardinal baseball game. Judy remembers Aunt Norma sending her to Straub's grocery store to buy a tomato. It killed Judy to pay a whole 50 cents for that one tomato!

Carl Edward (our father) was named after his sponsor, Uncle Edward Guebert. He was born in 1914 (died in 1990) and just like his brothers, had a nickname. His was "Chunky." He was baptized and confirmed at St. John Lutheran Church and graduated from Red Bud High School.

Carl as a small child

Carl went through some tough times as a child and in his youth. In 1922, his Grandma Guebert (Engel) died. Then in 1924, his mother (Rosa) died at age 56 after a stroke when Carl was only 10 years old. At that time, his father turned over the farm to Carl's brother Alfred (Bank). However, Carl, sister Norma, brother Lorenz (Honey), and his dad still lived on the farm with Bank and then with Bank's wife Norma and their children. That could not have always been an easy arrangement for any of them!

11-year-old Carl with his father (Charles), his Aunt Anita, and cousins Paul and Norman

From 1930–31 Carl's dad, Charles, made an extended trip to Canada and spent the winter with Carl's brother Dick. While in Canada, Charles wrote letters to Carl's sister Norma. In one of those letters we learned that 16-year-old Carl had scarlet fever and that the house had to be quarantined for four weeks. Carl was confined to the upstairs bedroom all that time! Carl's dad died in 1935 at age 67 from pneumonia and kidney disease when Carl was 21.

Carl worked for his brother Bank on the farm until he began farming on his own. He started renting the Kloepper farm northeast of Red Bud in 1939. We learned after Dad's death that during this time he had been a 4-H leader and that Arlin Obst (who later became Monroe County Farm Adviser) was in his club.

Carl and his siblings in 1952:
Walks, Dick, Bank, Vic, Honey, Norma, and Chunky!

ESTHER'S BACKGROUND

Esther Anne Christine Werner was born in St. Louis, Missouri, on April 6, 1918, to Ernst Alfred and Catherine (Katie) Garlich Gaebe Werner. Ernst Alfred Werner was born March 30, 1873, in Skaszchen, near Dresden, Germany. He had emigrated from Germany in 1884 when he was 11 years old. All of his brothers had the first name of Ernst which was their father's middle name, so they all were known by their middle names. So, from now on, we will refer to them by their middle names. Alfred married Anna Petri in 1908. She died just 8 months later after gall bladder surgery. Alfred was quite an accomplished business man as revealed in the following article that appeared in Volume II, pages 490–492 in the 1911 edition of *History of St. Louis* by William L. Thomas.

ERNST ALFRED WERNER.

There is always a measure of satisfaction in noting the progress made by self-reliant characters—men who, not depending upon higher educational advantages nor upon assistance from others, but rather upon their own resources, push onward with patience and perseverance, guided by a fixed purpose which they finally reach, and the attainment of which brings them to a high degree of prosperity. Such careers are possible in every walk of life, and it makes no difference what vocation one follows—if he desires success and expects to accomplish it he must be possessed of certain qualifications, without which he will eventually prove a failure in anything he may undertake. These qualifications are willingness to work, patience and perseverance.

Such a career has been that of Ernst Alfred Werner, senior partner of the firm of A. & H. Werner, horticulturists of wide reputation in St. Louis county. He is one of Germany's contributions to the citizenship of the United States, his birth occurring in Dresden, Saxony, Germany, on the 30th of March, 1873, a son of Gottlieb E. and Marie (Hillmann) Werner, also natives of Saxony. The former was born in 1848 and the latter in 1849, and on the 1st of August, 1869, they were united in marriage. The father was reared to farm work, but on reaching man's estate served an apprenticeship at the capenter's trade, and followed that occupation in his native country until building operations became slack in the fatherland.

In 1883 Gottlieb E. Werner came alone to the United States to seek a fortune and make a home for his family in the new world. He first located in Little Rock, Arkansas, where he attempted to follow his trade, but the radical difference between the building business in this country and in Germany made success in that line impossible and in January, 1884, he came to St. Louis county, Missouri, where he rented a farm in Central township, near the present location of his sons. His first intention was to carry on agricultural pursuits in connection with his building operations, but later he gave his entire attention to his farming and truck-raising interests. In September, 1884, his wife and children joined him in this country and the family home was established upon the farm which the father rented in that year. In 1892 he returned to Germany to attend the golden wedding of his parents, and while in the old country passed away, his death occurring on the 30th of October, 1892. His wife died on the 11th of February, 1907.

Ernst Alfred Werner spent the first eleven years of his life in the fatherland, and his education, which had been begun in the schools of that country, was completed in St. Louis county. He had been reared to farm work, early becoming acquainted with the tasks that fall to the lot of the country lad, and thus, upon the death of his father, in 1892, he was well equipped to assume charge of the work of the home farm, which he operated for his widowed mother until 1901. In that year in connection with his brother Hugo he purchased the present business location of the firm of A. & H. Werner, in Central township, and turned his attention to the truck-raising business. This tract, consisting of ten and a third acres, was in a badly neglected condition, upon which no crops could be raised, and with characteristic energy the brothers at once applied themselves to its improvement, enriching the soil by fertilization. They soon found, however, that the ground was too wet for the successful cultivation of truck gardens, and with undiminished enthusiasm they turned their attention to the horticultural business, their initial step being the growing of carnations. Almost from the first the undertaking proved successful, and from a very humble beginning it has since developed into one of the largest and best known horticultural enterprises in this section of the state.

The firm, known as A. & H. Werner, operate five greenhouses, four of which are one hundred and sixty-two feet by thirty-four feet and one one hundred and fifteen by eighteen feet, located on Spring avenue, between Olive street road and Page avenue, Clayton. They are wholesale growers of cut flowers and cuttings and make a specialty of roses and carnations. They have greatly developed the rose-growing industry here, in which line many other florists have failed, and have clearly demonstrated the fact that rose cultivation can be made a success in this locality. They are the largest rose growers in the vicinity of St. Louis, having on hand at the present time some fifty-three hundred rose plants.

Ernst Alfred Werner has also gained a reputation throughout this community as a man of considerable mechanical genius, being an expert in all kinds of wood and iron work. He first became interested in mechanical pursuits in 1894, when the business of truck-growing being slack, he began learning the blacksmith's trade, working in the shop of George Riegert on Olive street road. But business was not particularly good along that line of activity either, and after about six weeks he became restless and left the shop. This short apprenticeship, however, served to awaken the talent which had laid dormant in the young man, and since that time he has done considerable work along

492 HISTORY OF ST. LOUIS COUNTY

mechanical lines. In 1896 he erected some greenhouses for C. C. Sanders, and later was employed for a time by the Fairbanks-Morse people to build tanks and pumps. He designed and built all of his own greenhouses, and no better nor more modern structures of this kind can be found in the country. He has also done all of his own wagon repairing and horseshoeing for many years.

Mr. Werner was married on the 7th of October, 1908, to Miss Anna Petri, who, however, passed away eight months later. On the 18th of September, 1910, he was again married, his second union being with Mrs. Catharine Gaebe, who in her maidenhood was Miss Garlich. Mrs. and Mrs. Werner are members of the Lutheran church, in the work of which they are deeply and helpfully interested, and in his political belief Mr. Werner is a republican. He has never sought nor desired office, however, as a reward for party fealty, preferring to concentrate his entire energies upon the conduct of his business interests.

In his career he has ever manifested those sturdy qualities which are recognized as the birthright of the German race, and with these characteristics he has happily combined the spirit of progress and enterprise which is the dominant feature of the American race. He truly deserves the proud title of a self-made man, for whatever success he enjoys today is the direct result of his own efforts. On the strength of his own resources through the constant application of perseverance, coupled with hard work and careful management, he has utilized his energies to excellent advantage and has subserved them to such a degree that he has been able to rise in the business world from a humble position to his present high degree of prosperity.

Catherine Louise (Katie) Garlich, daughter of John Rudolph Garlich and Clara Maria Nienhueser, was born February 13, 1880, in Plum Hill, Illinois. She had married August Heinrich Wilhelm (William) Gaebe, son of Herman Heinrich Wilhelm Gaebe and Dorothea Detring, on September 23, 1903, in Addieville, Illinois, and moved with him to farm in New Salem, North Dakota. William died from diabetes complications on January 19, 1909, at age 26. This left 28-year-old Katie as a widow with 2 little boys: 4-year-old Ray and 2-year-old Hugo. She then moved back to the St. Louis area.

Katie met Alfred at a picnic at Grace Lutheran Church in Wellston, Missouri, where Alfred's band was playing. (Alfred played the trumpet in the Immanuel Lutheran Church, Olivette, MO, band.) Alfred won a friendship quilt that was raffled off at the picnic. Someone remarked that it was a pity that a bachelor had won the quilt. Little did anyone know that his future wife was at the same picnic! Alfred and Katie were married in 1910 at Immanuel Lutheran Church, (the same church where Judy and Kathy are now members!) and were blessed with four more children: Florence, Walter, Martha, and Esther.

Immanuel Lutheran Church Band (1898)

Back Row: Henry Kieffer, Alfred Werner, C. Wunnenberg, Julius Moeller, Geo. Kieffer, Hugo Werner, J. Fey, J. Wunnenberg, Second Row: Geo. Moeller, Geo. Hecht, Christ Fey, Theo. Wiebracht, Carl Mueller, Front Row: Wm. Moeller, Hy. Steinmeyer, Frank Mueller, Henry Stratmann.

Alfred and Katie in 1932

UNIQUE QUILT-- Esther Guebert of Red Bud shows a quilt that she inherited from her parents, the Alfred Werners, which has history behind it. The quilt was won by her father in 1910 at Grace Lutheran Church in Wellston, Mo. at a pincic. He had a band and played at church picnics. He, a recent widower, also met his second wife that day. The quilt has 12 circles with 20 names embroidered inside. A donation of 10 cents was asked from everyone whose name appeared. The quilt was on display at the Evening Guild's quilt lecture.

Esther with the quilt her father won. The quilt became the possession of Aunt Martha Werner, until her death, when Aunt Martha's daughters, JoAnn and Doris, presented it to Judy.

Katie had a difficult life. Besides being widowed at an early age, she dealt with severe rheumatoid arthritis. She actually became crippled because of it and was confined to a wheel chair around 1928 for 12 years (when Esther was only ten years old). Esther's father fashioned a wheel chair for Katie by taking a kitchen chair and putting wheels under it. Esther remembered her mom Katie making bead purses and some small wooden projects. Despite having severe arthritis, Katie kept her faith and a positive attitude and was an inspiration to others as shown in an essay which her niece Mary Jo (Joey) Garlich wrote about her.

Mary Jo was married to William and Anne Garlich's son Alfred Garlich. William was Katie's brother, and Anne was one of Esther's sponsors. During the two day visit that Mary Jo writes about, Mary Jo actually went with our mother Esther to St. Louis to buy fabric for Esther's wedding dress at Famous-Barr. They determined taffeta looked best on Mom, rather than shiny satin. She ended up buying several remnants to make her wedding dress. Eventually, she had to burn the dress as the taffeta didn't hold up. The two women always corresponded at Christmas-time but never saw each other again. Next is the essay that Mary Jo wrote about Katie.

Love Beyond Pain
By Mary Jo Garlich

Many years have passed since our visit to her, but she still remains in my memory as that unforgettable person.

Her name was Katherine Werner, but to many she was known as Aunt Katie, or simply as Katie.

On our way back from our wedding trip my husband and I decided to stop in Saint Louis and visit Aunt Katie and her family. On our arrival we were met at the door by her children. She was waiting for us in her wheel chair.

Her brown hair was pulled up into a little bun on top of her head, and her slender small featured face was reminiscent of a once beautiful woman. Aunt Katie's most outstanding feature was her eyes. They were large and beautiful. It was as though she had an inner glow. It was her expression that captivated me. Later I was to learn the meaning of that expression.

I had been warned by my husband not to clasp Aunt Katie's hand when he introduced me. "Even a touch is very painful to her," he said.

At the time of our visit she had been in a wheel chair for twelve years. She could move no part of her body.

Now came a pleasant time of getting acquainted, and renewing the relationship between Aunt and cousins. This was my first meeting with the family.

Aunt Katie's last trip had been a visit to Minnesota with her husband, and four youngest children, then in their teens. This was in 1927. My husband's last meeting with her had been on this occasion.

There was no sign from her that she was in pain, and no comment was made about her crippling arthritic condition. She would not have wished that.

I am sure we reminisced a little about old times, but Aunt Katie lived very much in the present. She was interested in what you were doing, and what was going on around her.

That night when her daughter Martha lifted her out of her wheel chair and into bed, I heard Aunt Katie cry out in pain.

On the following morning, during the conversation, she told us that she had a mission in life. When we asked what it was, she said that she was able to give cheer and encouragement to the people who came to see her.

More than anything else it was this statement by the most unassuming of persons, that made her so memorable to me.

Aunt Katie asked us if we would like to see the greenhouse. "I can't go out there with you," she said, "but the boys will show you around."

Two sons, Raymond and Walter had been in charge of the family business since the death of their father six years earlier. The greenhouse was next to the big brick house which was home to Aunt Katie and four of her adult children.

The flowers were beautiful and Raymond explained that they were specializing in carnations.

I asked Martha and Esther how they managed caring for their mother, and working. "We take turns caring for her while the other one of us works, and sometimes we are both needed here," they said.

The love and co-operation between family members was evident.

Aunt Katie had a canary that she enjoyed very much. At first it had bothered her to see it caged. She also liked to watch the birds in the yard. In the summer she could sit on the porch and watch whatever was going on outside.

It had been a pleasant visit, but we knew that we should start our trip home, so we said our good-byes.

That was an eventful year. First there was our visit to Saint Louis in February. Walter's marriage took place in March, and Esther was married in April.

Aunt Katie lived just long enough to attend the wedding of her youngest daughter. Kind hands wiped the tears of joy from her eyes. Soon after her sixtieth birthday she left this earth plane.

We were glad that we had seen her. I had heard so much about this very special lady.

Not long ago I made a phone call to Saint Louis and contacted Martha.

"I have always wanted to write something about your mother," I told her, "but all I have to go on is a visit of two days. In that two days she taught me some valuable lessons."

Martha was a little surprised to hear that after forty-four years my memory of her mother was still vivid.

"Am I correct, Martha, in saying that your mother never complained?" I asked.

"Yes, that is true," the answer came back. It was one painful day following another, and during the last twelve years of her life she was completely helpless.

"Is it true that her faith in the Lord never wavered?" I questioned.

"Yes," said Martha, "it seemed that she lived on to show her family and friends how strong faith can be. She truly put all her trust in the Lord Jesus, and was content and thankful for what she had."

Martha recalled having met one of Aunt Katie's old friends one day. "I remembered your mother's birthday today," she said. "Katie was always so cheerful and content."

Another old friend said that Katie was the best pal she ever had.

Other friends remarked that she was special and caring.

When the pastor felt depressed, he knew just where to go. He went to Katie's house to get that boost that only she could give.

"What was his problem?" I asked. "Was he, too, in need of her counsel?"

"No, it wasn't that," the answer came back. "You know how it is with a minister. He visits the sick and also listens to all the gripes and grumbles. Some of the complaints are petty. In the case of my mother, she gave thanks for her good eyes and ears. She gave thanks that her mind was clear and her speech intact."

I am sure that she also gave thanks for the four children who cared for her so lovingly. (The above is my own thought—not Martha's.)

"Did you ever feel resentful about having to devote your time to your mother?" I asked.

"No," Martha said, "there was much love there. I am sure we all felt the same way. We wanted to do this for her."

"How did Aunt Katie amuse herself?" I asked.

"As long as possible she sewed our dresses," said Martha. "A dear friend brought her all the quilt patterns, so she sewed and quilted them. She also loved doing medira or eyelet embroidery. We girls cut out quilt pieces and pinned them together. When her hands could no longer sew them, she turned to the sewing machine. Another hobby was drawing house plans for people. In time, she got so she could do none of these things, and her pastime was sitting on the porch. We were in the country and not too much was going on, but there were always the birds to watch. We would take Mother to a church picnic or to the park and she could sit in the car and visit with friends."

"There was the night," said Martha, "that Mother awoke to see Raymond moving around with a flashlight. Mother knew that something was seriously wrong. Her heart was pounding, but she lay very quiet pretending to be asleep, and trying to keep calm. She knew that it would only be a bother asking questions. Suddenly Ray was anxious to get Esther to the doctor. Mother prayed that whatever was wrong would be righted."

"As soon as he could, Ray let the family know that Esther had gotten to the doctor on time. The culprit was a black widow spider."

This story gives an insight into Aunt Katie's character. As always, she thought of herself last.

Martha's description of her mother is that she was humble, sincere and thoughtful. She was witty, too.

"She loved people and was interested in everything they did, be it ever so simple. Her health problems were not on her list of things to talk about."

I have no idea what the extent of Aunt Katie's formal education was. Her wisdom was born of love, and that love very surely came from above, and shone on her sweet kind face.

"I don't know why, but God wants me here for a reason," was one of Aunt Katie's quotes.

Now all who knew Aunt Katie Werner could easily guess what that reason was. Her mission in life was definitely accomplished.

As long as I live, I shall never forget her. She remains in the memory of all the loved ones she has left behind.

Esther's oldest half brother Ray Gaebe was born in 1904 (died in 1977) and in 1950 married Vicki Zircher Bruder, who was widowed with one daughter, Vickie Ann. Ray worked in the family greenhouse business. Uncle Ray and Aunt Vicki would come to the farm to visit and would always bring a bouquet of snapdragons and carnations. Uncle Ray was a tall, soft-spoken, gentle man (much like his godson, our brother Steve). His hands were all misshapen and gnarly from rheumatoid arthritis, much like Grandma Katie's probably were. We loved to listen to Aunt Vicki in her Hungarian accent.

Next was half brother Hugo Gaebe born in 1906 (died in 1984). He married Mary Howard, and they had 2 daughters,

Jessie and Mary Grace. Mary was rather introverted, so we did not visit with them very often. Uncle Hugo also had the gnarly hands from rheumatoid arthritis, but so far, the disease has seemed to escape the rest of the family.

Sister Florence (Judy's sponsor) was born in 1911 (died in 1992) and married Lutheran school teacher Hans Mueller in 1932. They lived in Michigan until the 1960s when they moved back to St. Louis. They had four daughters; Dorcas, Vera, and twins Mary and Martha. While in Michigan they often traveled to St. Louis to visit family and would occasionally stay with us. Their visits meant we were going to get hugs, which was a rarity for us! Once when Kathy was a teenager, the subject of her birthmark on her leg came up. Kathy was very self-conscious about it. Aunt Florence said not to worry, that when Kathy walked down the aisle in her wedding dress, no one would see it. She was right!

Next came Walter (John's sponsor), born in 1913, died in 2008. In 1936 he married Elsie Weiss, and they had three children. Walter worked in the family greenhouse business. We visited Uncle Walter and Aunt Elsie periodically. When we went to see them, Esther would always bring along a loaf of bread and hot dogs for dinner. Remember, we were a family of 10; that could be a bit daunting for mealtime! They were one of the few aunt and uncle couples with children close to our ages: Ellen, Alfred, and Claudia.

Martha (Lois's sponsor) was born in 1915, died in 2002. She worked for the Ross family, where Annette Meyr (Judy's future mother-in-law!) was a maid and server. Martha's job was cook and downstairs duty. She and Annette slept in a room above the garage. Mr. Ross was President Truman's press secretary and when Truman went to Washington, D.C., Martha went with him. After 4 years, she moved back to St. Louis, when Mr. Ross moved to Kansas City.

Through a friend from Perryville, Martha met Otto Schlichting, whom she married. Otto farmed near Wittenberg, Missouri. Going to visit them was always an adventure. They lived about an hour's drive away in the rolling hills of Perry County. We always made "barf bags" out of newspapers to take along if we were traveling to Wittenberg. Invariably, one of us would get sick. One memory Judy has is of her and Lois, ages 8 and 5, staying overnight with Uncle Otto and Aunt Martha. Uncle Otto took them to see an Abbot and Costello movie in Perryville. Otto laughed his head off, and Judy laughed at him! Aunt Martha was a great cook, and there would be delicious homemade bread and a huge plum coffeecake waiting for us when we arrived for a visit. Aunt Martha and Uncle Otto's children were close to Kathy in age with JoAnn being a year older and Doris a few years younger. They also had a son born in 1951 who died shortly after birth of hydrocephalus. Mom commented once that going to his funeral was very difficult, especially since she had a healthy baby (Kathy) who was just a month older.

<u>Esther</u> was the youngest, born at home on Spring Avenue (now North and South Road) in University City on April 6, 1918. She was named Esther Anne Christine after her 2 sponsors, a customary tradition then: her Aunt Anne Garlich and Aunt Christine (Tina) Werner. Esther had a nickname, too. Hers was "Eppie", which she acquired because brother Hugo had trouble saying Esther and called her "Eppie." She was baptized and confirmed at Immanuel Lutheran Church in Olivette, Missouri, where Judy and Kathy are now members.

Her best friend was Luletta Wittmaier Klingert (Tom's sponsor). Luletta always thought Mom was very smart and would sit next to her in school so Mom could help her. Mom always loved Luletta's beautiful red hair. Luletta tells us she always thought the Werners were rich, because Mom's family always had candy at their house. Luletta said the Klingerts were very poor and their large family lived in a shack where Monsanto (Bayer) is now. Mom did not have the opportunity to go beyond half of 9th grade as she was excused to help take care of her mother. Kathy asked her once if she had been able to go on for more schooling, what profession would she have chosen. She said she would have been a nurse. Even though Esther was not highly educated, she was very intelligent and very talented. Her thriftiness and love of sewing trickled down to all three of us girls.

Esther as a small child

Esther's Confirmation picture

Werner home in University City, Missouri

For several years, Esther's Grandma Garlich (Katie's mother) lived with Esther's family in the winter months and with Esther's Uncle Bill and Aunt Anne (Esther's godparents) in Minnesota during the summer. Grandma Garlich had her own bedroom upstairs in the University City home where she would sit in her cane rocking chair and knit lace. She was a petite lady and usually wore a long black skirt and a dark or white blouse. Grandma Garlich would get out empty tin tobacco cans and cigar boxes she had saved from Grandpa and play building blocks with Esther. Esther never knew her Grandpa Garlich; she thinks he committed suicide in 1903. Evidently, he had a hard life, and Grandma Garlich never talked much about him. Grandma Garlich died in 1937 at age 85.

Esther's Grandma Garlich in the 1890s

Apple butter making in University City in the 1920s.
Aunt Tina Werner on left, Grandma Garlich next to her,
and other two unknown

Grandma Garlich and Katie with Florence and daughters
Dorcas and Vera in 1936

Esther once recalled a family trip she made with her parents, brother Walter, and sister Martha in 1927 in their "Durante." They went to Pipestone, Minnesota, to visit the Garlichs, Katie's brother and family and then to North Dakota to visit the Gaebes. On the way home somewhere in Illinois, the car caught on fire. Alfred (Esther's father) called the local Lutheran pastor who found a place for them to stay overnight. The car was not drivable so they came home on the train.

Alfred and his brother Hugo, ran the "A. and H. Werner Company," specializing in roses and carnations. Mom remembers her dad designing and building a large cart on tracks to the greenhouses in order to move and change the dirt for the flower beds in the greenhouses. Other florists and greenhouse businesses copied his plans. Later, chemicals were used, and the soil didn't need changing.

During the depression people weren't buying flowers. The greenhouses were located in Clayton on Spring Street, now known as North and South Road in University City. Evidently, in the early 1930s the company went through a bankruptcy and at the same time was forced out of Clayton because of the pond and also the large sooty smoke stack. Supposedly, Hugo thought they should move to the St. Charles, Missouri, area, but they ended up in Millstadt, Illinois.

Hugo moved his family first in March of 1932 to the house known to us as Aunt Tina's. Alfred's family moved in 1933, first to a house in town. In 1934 they moved next door to Hugo and Tina to the white "Skaer" house. Unfortunately, Alfred died of uremic poisoning that same year at the age of 61. Katie and family moved in with Hugo and Tina, staying in the west half (2 rooms) of the house while Hugo's family had the east half. The boys slept upstairs, with one boy always sleeping in the greenhouse for maintenance purposes.

Hugo Werner (Alfred's brother) had married Christine (Tina), who was Esther's godmother. In early years Esther remembers brushing Tina's very long hair, which Tina usually wore in a bun or braided and wrapped around her head. Hugo died in 1949, so we do not remember him. However, we do remember visiting Great Aunt Tina. She and Hugo had 3 sons: Ernest (David's sponsor), Herbert (Roger's sponsor), and Robert. Ernie was just seven months younger than Esther and was like another brother to her and an uncle to us. Esther told stories about Ernie: Ernie playing barber and cutting a hunk out of Esther's bangs and Esther being talked into carrying his schoolbooks for him. Ernie was a tinkerer and inventor, probably much like Esther's dad. Ernie's wife Doris was always welcoming and like an aunt to us. They had one daughter, Jeanette, the same age as Kathy. Ernie died January 13, 2013.

After Esther's family moved to Millstadt, Martha took care of Katie while 16-year-old Esther started working in a garment factory in town, sewing lingerie and earning 32 cents an hour! While working at the factory, Esther got sick from heat exhaustion and had to take off work for two weeks. About that same time Martha went to work as a housekeeper in St. Louis. So, Esther quit the factory job to take care of her mother until her marriage in 1940. Sadly, Katie died of pneumonia just a week after Esther and Carl's wedding in 1940 at the age of 60.

Esther and her siblings, 1964:
Florence, Martha, Walter, Esther, Ray, and Hugo

CHAPTER 2

The Early Years—the 1940s

Walther League was a very active Lutheran social group for young people high school age and up into their 20s. Local Walther League groups often attended zone rallies and the like. Carl and Esther met at one of these gatherings. Their first date was a picnic double date to Forest Park in St. Louis with Harold Gielow and Ruth Lakeman. Carl and Esther were married April 14, 1940, at Trinity Lutheran Church in Millstadt, Illinois. Rev. Deichman was the preacher, and teacher Sommerfeldt from Red Bud was the organist. The attendants were Aunt Martha, Aunt Norma, Uncle Walter, and Buddy Sauer (Dad's cousin and David's sponsor). The quartet of Alex Loesche, Luke Koch, Carl Burgdorf, and Barney Gielow sang. (Dad was a member of the St. John's octet.) Afterwards, they had a reception at their future home at Kloepper's place northeast of Red Bud, where Carl had started dairy farming on his own about one year prior to their marriage.

Katie, our grandma, did not attend the wedding but was at the reception. We assume she came to stay with Dad and Mom as there was no longer anyone else to care for her. She died just a week later with Mom at her side. She was only 60 years old; Mom thinks she had pneumonia.

Buddy Sauer, Walter Werner, Carl, Esther, Martha Werner, Norma Guebert — April 14, 1940

It must have been a hard adjustment for Mom, losing her mother and switching from city life to farm life, especially when the babies started coming! Remember, there were no grandparents to help out, and Dad (Carl) was not one to help out with the babies or the housework. However, Esther was very helpful with the farm work, particularly with the milking every morning and evening, until the boys were old enough to help.

In the fall of 1940, while Dad was hauling logs from the woods to the house, an axle broke on the wagon. Somehow, Dad's left arm was badly broken. Mom took him to the hospital in Red Bud. He was then taken to a St. Louis hospital to get it set. While it was in a cast, Dad continued to work and milk cows by hand.

The arm healed crooked, and we kids will always remember his swayed left arm from elbow to wrist.

Dad often commented that the Lord said, "Be fruitful and multiply," and that's what he and Mom did! Judy was the first of 8 to be born. Judith Catherine Guebert was born May 1, 1941. She was born with an extra finger on her left hand and an extra toe on each foot. The extra appendages were all removed while she was still in the hospital. The extra finger and toes were actually not a big surprise for Mom as polydactyly (extra fingers and toes) was a hereditary trait that her grandmother Clara Nienhueser Garlich (Katie's mother) had. From then on, whenever another child was born, Mom's first question was, "How many fingers and toes does the baby have?"—not whether the baby was a boy or girl! Mom mentioned that we may be descendants of the giants in 2 Sam. 21:20: *"There was a man of great stature, who had six fingers on each hand, and six toes on each foot, twenty-four in number, and he also was descended from the giants."* However, in the same breath, Mom said it may not be something to be proud of!

Katie was not born with any extra fingers or toes, but 3 of her children were: Hugo, Florence, and Walter. One of Florence's children and one of Walter's children had extras. Mom did not have any extra fingers or toes, but 3 of us children did: Judy, Steve, and Tom. However, we are getting ahead of ourselves here!

Keep in mind, when each one of Dad and Mom's babies was born, Mom was in the hospital for a week. That was very common until the 1970s, as Judy was hospitalized a week after both Jodi and Randy were born ('64 and '66), but only a couple days when Jeremy was born in 1981. So, we're not sure how Dad managed with Mom being gone a week with each birth until Kathy came along. By that time, Judy was 10 1/2 and able to help at home.

Judy and Brownie at Kloepper's, 1942

Did you notice that sagging diaper in the photo? No Pampers back then! Cloth diapers had to be hand rinsed, soaked in bleach in a bucket or diaper pail, washed, hung to dry, and folded—a job in its own. This was the norm for most of our own children. There were rubber pants that we put on over the diaper. In the '60s and '70s, Tidy Didy diaper service was a thriving business *if* you could afford it. Judy had a shower gift of one month's diaper service, and it was quite a treat. They picked up the dirty cloth diapers and left a set of clean ones. Pampers were invented in the '50s but didn't become popular until the '80s. That's also when diaper services went out of business.

Our parents and grandparents were actually much more environmentally friendly than we are. They were masters of the 3 R's: reduce, reuse, and recycle before anybody ever heard of the phrase! Years ago, both seed and flour were purchased in big cloth bags with designs and when emptied, the cotton fabric was used to make dresses or towels. Ever heard of feed sack dresses? The first quilts were made from scraps of fabric. Vegetables from the garden were canned in glass jars, and homemade jelly was sealed with wax in glass jars. Those jars were then used over and over again. If you bought soda or milk, a deposit for the glass bottles was included in the price, and you got your deposit back, usually 5 cents per bottle, when you returned the bottles. Feather pillows were the norm. Mom made ours from geese that were owned by her cousins, the Isringhaus's from Plum Hill, Illinois. "Waste not, want not" was a common expression.

Judy and Kitty at Kloepper's, 1942

John Paul Guebert, Dad and Mom's second child, was born on July 19, 1942. Dad had fixed up the Kloepper farmstead with a milking parlor, thinking he was all set to stay there a while. However, in August of 1942 when John was just a month old, Kloepper's son decided he wanted to farm. So, Dad had to find another place. He and Mom then rented Hepps' place just outside of Hecker off Route 159.

Hepp's farmhouse, 1942–1946

Cattle and Barn at Hepp's

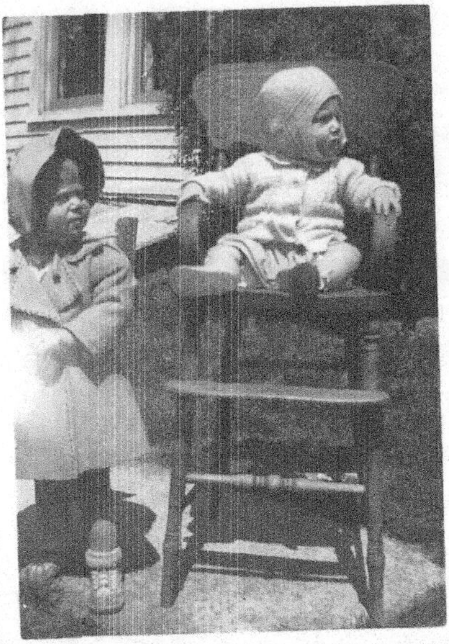

Judy and John at Hepp's

Dad and Judy at Hepp's, 1942

Judy and John with cows in background at Hepp's

On February 10, 1944, Lois Rose Guebert was born. Mom never made it to the hospital for her delivery; the doctor came to the house instead!

Lois, John, Judy, and Mom (pregnant with Steve), 1945

Lois, 1945

Lois and Judy, Summer 1945

Steven Ray Guebert was born July 21, 1945. Not only did he have extra toes, but one leg was a little shorter than the other leg. We do not know if this is a hereditary characteristic that goes along with polydactyly. His surgery to remove the extra toes was done when he was one, the same time Judy had corrective surgery on her left hand. This was also about the time Dad and Mom moved to the current Guebert farm! Judy recalls looking out the St. Louis hospital's 3rd or 4th floor window, watching Mom and Dad walk across the parking lot to their car, and feeling terribly sad and afraid. She and baby Steve were in a ward of eight cribs, and she'll never forget the strong, unpleasant smell!

Steve and Lois, 1945

Judy and John, 1945

Judy remembers the Hepp place where the chickens had free range. It was fun when they delivered the new baby chicks. They came in boxes with peepholes and were so soft and cute as the dickens! Judy loved to feed the chickens, spreading the feed in a spiral pattern on the ground; the chickens lined up in that spiral pattern to eat! When she was 5, a mean old rooster chased her through the yard. She quickly climbed a fence, but the rooster won—he flew up and pecked her on the arm so hard she has a scar to this day!

By 1946, Dad had saved up enough money to put a down payment on a farm of his own. He and Mom purchased the Miner farm at a tax sale at the Monroe County Courthouse for $100 an acre. So, in August of 1946 Carl and Esther and the four kids (with number five on the way) moved to the two-story farmhouse on the 157-acre Miner farm located just two miles south of Hecker. The dining room was originally a log cabin, probably built in the early 1800s. Some of the large logs are still visible from the basement under the dining room. The two-story brick portion was built around 1900. This house, as well as three others nearby, was built by a Mr. McQuillan for his daughters. Mr. McQuillan had owned a whole section of land (640 acres). He evidently sold our house after about ten years to the Roscoes (a daughter and son-in-law). They in turn sold it to the Miners.

Judy and John, 1946

Judy remembers a lady showing the house and pointing out the lilac bush where the Easter bunny always hid eggs! There was running water in the kitchen, but no indoor bathroom at the time. Having to use an outhouse was still pretty common in the 1940s. However, one of the first things Dad did was to put in a bathroom. The upstairs rooms were not heated, too cold for the little ones, so they slept in the living room the first couple years.

Cousin Vera (Aunt Florence's daughter) shared memories of the move from the Hepp farm to the Miner farm. Eleven year-old Vera and her family were in the area for an extended visit in 1946, while her father, Uncle Hans, was taking some classes at Concordia Seminary in St. Louis. So, they were available to assist in the move. She remembers the Miner outhouse being much nicer than the one at Hepp's, as she says she "picked up a yeast infection, probably caused from the primitive conditions at Hepp's place!" Vera also remembers that initially we did not have an electric refrigerator. So, Esther would drive to Hecker every other day to buy a block of ice to put in the old-fashioned ice box. The block of ice was put on the front bumper, and Mom would drive carefully back to the farm!

The garage had originally been a summer kitchen. (Separate kitchens were used for cooking in the summer so that hot kitchen activities did not heat up the rest of the house.) Large shale stepping stones along the south side of the garage led to the outhouse to the east. A grape arbor served as a privacy fence in front of the outhouse.

No indoor bathroom also meant no modern-day bathtub either. So initially, water was heated on the stove and poured into a metal tub. Baths were a weekly event on Saturday nights so you were clean for Sunday morning church! The water was reused. You started with the least dirty kid and ended with the dirtiest one! This sounds gross now but was a common practice in the 1940s and even in the 1950s.

David Allen Guebert was born October 5, 1946. He gave everyone a scare as a baby when it was feared he had polio. His room (Mom and Dad's bedroom) was quarantined for a short time. Fortunately, that was not the case; it was not polio.

David's baby picture

Werner relatives, 1947: Back: L–R Aunt Elsie, Uncle Walter, Uncle Ray, Aunt Florence, Uncle Hans, Carl (Dad)
Middle: (Baby) David, Dorcas, Vera, Esther (Mom pregnant w/ Tom)
Front: Ellen, Judy, Martha, Mary, Lois, Johnny, Steve

The year after the move to the Miner farm (1947), Judy entered 1st grade at McQuillan public grade school, named after the original owners of our farm. She remembers being so scared her first day of school and Mom fixing braids across the top of her head while she sat on the kitchen stool that morning. By the way, none of us went to kindergarten; it was not required back then.

Judy ready for her 1st day of school, 1947
(Note the paper Stag beer lunch bag!)

McQuillan was one of the last one-room school houses in Monroe County. There were only 12 students in all eight grades. It was located on the county road north of our farm, on the exact spot where David's friend Omar Koester lived for a while. It's now called Roscoe Road. Dad made a path and built a bridge over the creek in the woods to cross over. Nowadays one wouldn't think of sending a 6-year-old through the field and woods alone to walk to school! There was a cloak room on each side of the front door; left side was for girls, right was for boys. John remembers the pot-bellied stove in the schoolhouse, the old outhouse, and at recess playing on the bank of the road with matchbox cars.

Judy loves to tell these stories. . . .

> When I was in first grade, Willa Mae was my classmate. Herbie flunked 2nd grade, so it was Willa Mae, Herbie, and myself as second graders. Herbie flunked again, so in third grade, it was back to Willa Mae and me again.
>
> With John being just a year younger, we were first and second graders. One day Johnny had an accident (in his pants), so Mrs. McDermott had me walk Johnny home for a change and then we walked back again.
>
> I was in third grade when Lois got so sick. It was the year she had one of her kidneys removed, due to a Wilms' tumor. I remember Mom picked us up from school and my teacher had a bunch of cards for Lois. They were beautiful doll cards with feathers in their hats. For some reason, I remember feeling jealous, because she was my teacher and Lois wasn't even in school yet.

Thomas Carl Guebert came along on September 19, 1947, less than a year after David was born! Tom had an extra finger that was removed when he was born.

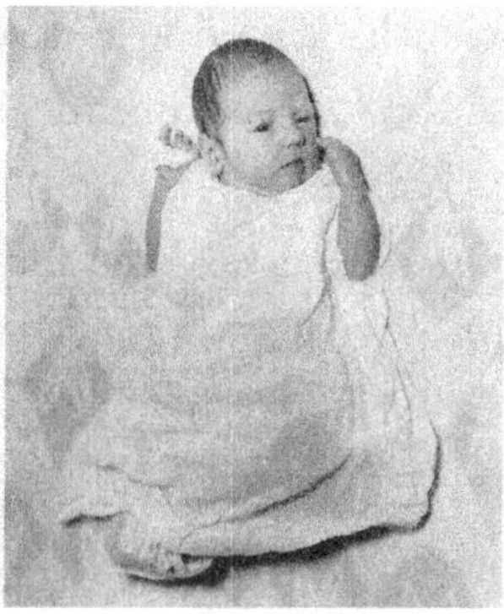

Tom's baby picture, 1947

Judy in front of chicken house before 2nd half was added, 1947

Judy, John, Lois, Steve, in front of feed-house, 1947.
Note, the barn on the left is the barn that burned down in '55.

Tom and Lois (in a swimming suit), August 1948

Tom, Lois, David, John, Judy, and Steve—Summer fun!

Lois, Tom, Steve, David, 1948

Boy number 5, Roger Dale Guebert, was born October 30, 1948. Carl had plenty of future farmhands now!

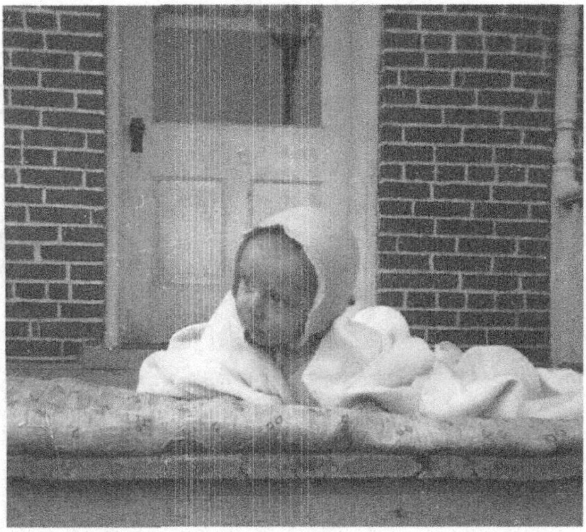

Baby picture of Roger, 1948

Steve, Lois, and Judy, Christmas 1948

Carl and Lois, Christmas 1948

Roger and Esther, 1949

Lois, John, David, Steve, Judy, Tom, 1949

After being a sickly child for several years, Lois was diagnosed with a cancerous Wilms' tumor in one of her kidneys in 1949. The kidney was removed, and she did fine the rest of her life with just one kidney. To this day, we do not know how Mom and Dad managed with Judy and John in 2nd and 3rd grades, Lois in the hospital, and four toddler boys/babies at home! There must have been a babysitter we don't know about!

The first few years when the kids were still too small to help out on the farm, Dad hired a farm hand, Henry. Henry lived with us and slept in the small middle bedroom upstairs. He was an elderly man, very quiet, and kept to himself. Dad employed him until Henry could no longer work and then helped move him into the county nursing home. We remember Henry spending most of his weekly paycheck at the nearby tavern and coming home inebriated, but harmless, and sleeping it off.

Esther had some help with the babies. Uncle Walks and Aunt Alma's daughter, Rosie, helped out on occasion. One summer, probably around 1946, when Rosie would have been 13 years old, Dad paid her a whole $5.00. She went along to the Hecker homecoming and accidentally dropped her wallet with the $5.00 in it down the pit in the outhouse! $5.00 was a big deal back then, not something to just dismiss, so someone (we don't know who!) fished it out for her!

You might be wondering where everyone slept. Mom and Dad had the master bedroom on the first floor where there was also a crib for the youngest child. The crib was located where the current closet is. The upstairs bedroom on the east side was the boys' room, and at one time all 5 boys slept in there! There were 2 double beds and a twin bed in that room. The upstairs bedroom on the west side was the girls' room; it had a double bed and a twin bed in it. When it was time for one of the boys to get up, Mom would call (or yell if no answer—the only time we ever remember her yelling!) from the bottom of the stairs. She would go down the line, calling all the boys' names even after the oldest ones were no longer at home! She was our alarm clock.

None of the bedrooms were built with closets, other than the little closet above the stairs in the small middle bedroom. Mom and Dad later built the closets in the other three bedrooms. When Henry left, John took over the middle room. When John left for the Navy, it was Steve's room, then John's again for a short time when he came back from the Navy. Later on, Kathy even claimed that room and used it for her 4-H room improvement project.

The original house had a small square open porch on the front and a big L-shaped screened in porch along the north side of the master bedroom and the east side of the kitchen. In order to fit in a bathroom, the bend of the L was walled in, which left two separate porches. This square room was large enough for a bathtub, sink, and toilet. However, there were two doors into it, one off the dining room and one off the hallway. Needless to say, it was difficult to get much privacy or time to yourself in the bathroom!

Originally, the washing machine (a wringer type) was in the basement, and the "dryer" was the clothes line outside or the clothes line in the basement! That never dampened Mom's spirits, though. Kathy remembers Mom lifting the wet clothes out of the washing machine with a short broom handle, then running them through the rollers to squeeze out the water, and hanging them up to dry outside. Several wash lines ran from the back of the garage to the outhouse. The pole for propping the line was made from a long, forked tree limb. During all this time, she would be humming as she worked.

The original source of heat for the house was coal. Dad later switched to wood, because of the soot issue and a wood source on our property. Kathy remembers Mom having to get up during the middle of the night to stoke the furnace with more wood to keep the furnace running. There was a special trap door in the furnace room for the wood. We eventually switched to a modern furnace that used natural gas. The upstairs bedrooms were not heated, so we would rush downstairs with our clothes and fight over a heat register to get dressed!

There was no air conditioning, either. AC wasn't installed until some time in the early 1970s. We begged for it for years! That meant the windows were usually open in the summer. That also meant little tiny bugs (oats bugs) that were attracted to the house lights would come in through the window screens. In the morning all these tiny bugs, now dead, would be lying on the dining table - ugh! We also made good use of fly swatters. We can still see Mom with a fly swatter in her hand during the summer months.

Our mother was very versatile in many ways. She would even tackle wallpapering and painting the big tall rooms in that house. That was before pre-pasted wallpaper.

Throughout the years, other remodeling projects were done to the house. In 1962, Dad hired Buddy Sauer (Carl's cousin and best man for their wedding, and David's Godparent) to build new cabinets in the kitchen. He also remodeled the bathroom, taking the porch area next to the master bedroom and converting it into a spacious three-room bathroom. One room housed the bathtub (now with a shower attachment), another the toilet, and another the sink area. The little square area that had been the bathroom was now a pass-through with a built-in closet. The other section of the porch was closed off also, as an entry area and a mudroom.

The mudroom created a place for farm work clothes and also housed a sink and later on, the washer and dryer which had been in the basement.

In 1965 the little porch on the front of the house was taken off in place of a large porch, complete with pillars. It now looked like we had a southern mansion!

House with small porch

House with pillars

Chronologically we have gotten ahead of ourselves, so let's move on to chapter 3!

CHAPTER 3

The 1950s

It is 1950, and Carl and Esther have 7 children with the oldest only 9 years old!

Tom, Judy, Roger, John, Lois, Steve, and David, 1950

All 7 kids with Aunt Florence and Esther, 1950

David, John, Steve, Tom, Lois, Judy, and Roger
with chicken house in background, 1950

In 1950 when Lois was ready to start first grade and now with three children in grade school, Dad and Mom decided to start driving them to Red Bud to attend St. John's Lutheran

School. Even though we were only 5 miles from Red Bud, we were in the Waterloo School District, which meant no bus service to Red Bud. That twice a day drive to Red Bud continued for 15 years until 1965 when Kathy graduated from the 8th grade.

We had memorable teachers in grade school. Lois had Neva Luebkeman for 1st and 2nd grades while Judy and John had cousin Evelyn (Guebert) Muench for 3rd, 4th, and 5th grades and then Mr. Sommerfelt for 6th, 7th, and 8th grades. Later, it was Miss Seager for 1st grade, on whom Tom had a crush, as did all the boys! It was a big disappointment when she told them she would not be back as she was getting married. Later on, when Kathy came along, it was Mrs. Kuring for 1st and 2nd grades. Third grade was Mr. Gutknecht, and 4th grade was Mr. Prothe. Mr. Prothe would actually dig his thumb into your shoulder inside the collar bone if you misbehaved, and you needed to be sure to eat everything at lunch because you had to turn the plate over to prove it! Mr. Prothe had an "old time" day each year when you had to be very still and not get out of line or you were punished by having your hands smacked with a ruler. However, he also had a special day when we got ice cream treats! He actually died of a heart attack when Kathy was in 4th grade; maybe it was the forks full of butter Kathy remembers him eating! Mrs. Stellhorn, 5th and 6th grade teacher, was a favorite of Steve's and the rest of us. She had us do research notebooks on a state, then a country. We also had to make relief maps on poster board of the Rockies and other mountain ranges out of flour and paste. John was the first one to have Mr. Snyder, and all the rest of us were privileged to have him as well. Mr. Snyder was Tom's favorite teacher because he was young and coached sports. He was the reason Tom decided to be a teacher/coach. Tom remembers walking in line from school to the movie theater two different times, once for an Abe Lincoln movie where we all received a shiny new penny and again for the movie "Exodus" with Charlton Heston as Moses. He also remembers one of his favorite lunches at school was on Thursdays, when we always had chili soup and a homemade donut—the big ones!!! They were the best!

Below is a picture of the old St. John Lutheran Elementary School just before it was demolished and a picture of the new school which replaced it. The old school was a huge two-story brick structure with tall ceilings and a lunchroom in the basement. A modern one-story brick structure which included a gymnasium was built in 1961, in time for Roger's last year of grade school and Kathy's last four years.

One more baby was born December 24, 1951, Kathryn Anne Guebert. Finally another girl after four boys in a row! Judy, age 10½, remembers it well, as it was Christmas Eve. She had six kids to get ready for the Christmas Eve service, including fixing seven-year-old Lois's hair! Judy remembers getting a toy typewriter for Christmas that year and typing a letter to Mom while she was in the hospital. She had to ask Dad how to spell hospital. Lois remembers Kathy's baptism in the public school gymnasium, as the church was being remodeled. Judy remembers playing a favorite baptism hymn on the piano at home, hoping it would be sung at her baptism, and it was!

About that time the family got our first bicycle. Dad tried it out first. We watched and laughed from the kitchen window because he slipped on the ice and fell, which probably wasn't anything to laugh at!

Kathy, 1952

John and Kathy with the family bicycle

Carl with all 8 children, 1952

The 3 sisters, 1952—Note the brooder house and gas tank on top of railroad ties.

Being number 8 with a three-year break in between sometimes left Kathy wondering if maybe Mom would have liked to have stopped with number seven. Years later, as Mom was leaving after helping Kathy with newborn granddaughter Annie, Mom turned around and told Kathy she was happy that Kathy had a little girl, because she was sure happy when Kathy was born!

There were some advantages to being the youngest in the family. Some would call it being spoiled! When all the other kids were in school, Esther would take Kathy with her to Home Extension day meetings in Waterloo. Kathy remembers quietly playing with paper dolls in the second-floor meeting room of the Farm Bureau building while the ladies went about their business. Kathy also often accompanied Esther on shopping trips to Red Bud, and if she was really good, sometimes Kathy would get a 6-cent fudge bar from Schreiber's general store. Schreiber's is where we girls spent a lot of time choosing patterns and fabric for our sewing projects!

Even with farm life and eight children, Dad and Mom did not shrug from church and community involvement. Both were very service minded and active in the church. Dad served numerous positions on the Church Council and Men's Club, sang in the choir, and played dart-ball. He was also Waterloo School Board President for many years, awarding diplomas to us kids. Mom served numerous positions with Evening Guild all through the years. She was very active with Home Extension and 4-H in Monroe County, serving as a 4-H leader for over nine years.

There was many a night Judy served as babysitter without even realizing it! Speaking of babysitting, Judy did major babysitting when she was only 12 years old. With Aunt Florence's encouragement, Mom decided she needed a break. She went to visit her sister, traveling by train to Grand Rapids, Michigan in 1953. Surely, she needed the break, but Judy must have needed one after that too! Mom took 1½-year-old Kathy along with her, leaving Judy with six kids to take care of! Judy remembers complaining to Dad that she couldn't get any respect from those kids! One of those nights, she cooked a beef roast in the domed roasting pan Mom always used that had pieces of aluminum foil stuck where the handle should have been!

Esther and Kathy in Michigan, 1953

Esther and Kathy in Michigan, 1953

Judy didn't realize until she had kids of her own that she was the family babysitter when Mom and Dad both had evening meetings. She would sit at the kitchen table after all the kids were in bed and watch for the car lights to turn off Route 159, into our lane. By the way, back in the '40s and '50s, the highway was called the "hard-road."

All eight children in front of garage, 1953

Pitching in was part of farm life. When Judy was 16 and had a driver's license, she drove the truck full of wheat up to the Hecker grain elevator. Then another time she had to drive to the quarry on the New Athens road to get a load of rock. John went along in case she had trouble. He was actually the better driver but didn't have his license yet! Judy still does not know how she made it out of the bottom of the quarry pit, up the hill, and home with that stick shift truck, which swayed back and forth on Highway 159!

The boys helped with chores, field work, and milking the cows as soon as they were old enough to do so. They built up their muscles at hay and straw baling time, throwing and stacking 40-pound bales onto the wagon and into the barn.

There were also hogs to feed and eggs to collect and candle. We would gently tap two eggs together and listen to the sound that it made to tell whether or not there were any cracks in the eggs. (Judy stills taps her eggs together before cooking!) David remembers once dropping a whole basket of eggs (all broke of course!) as he was carrying the basket down the basement cellar steps. Roger remembers this happening to him when he was carrying two baskets of eggs! A whole basket of eggs broke another time, in the car as Esther was driving us kids to school, taking the eggs with her to be delivered to the grocery store in Red Bud! Judy said she was holding the basket, but she doesn't recall how the eggs broke!

There were even sheep (an FFA project of John's) and bandy chickens (a 4-H project of Roger's) for a while. The girls mainly helped in the house with the cooking and the many dirty dishes. Remember, there were 10 plates, 10 forks, 10 knives, 10 cups, and the accompanying pots and pans and no electric dishwasher. Dishwashing was a nightmare! Judy and Lois often quarreled over who was washing and who was drying!

Despite the work, there were fun times too.

- We liked to ride on the fender of the old Allis Chalmers tractor for hours, and we sometimes would watch that Dad did not hit the corn or beans with the tractor tires.

- We played outside a lot, making mud pies with our little metal cake pans and making roadways for "Follow the Leader" by sweeping paths with a broom through the gravel.
- We made roads for the matchbox cars by tearing off clumps of green moss from the tree trunks for lawns, and had little houses that Judy made out of cereal boxes.

Roger, David, and Tom Playing with matchbox cars, 1958

- We had tunnels and even hidden rooms amongst the bales in the hay barn.
- The rebar in the hay barn served as perfect bars for acrobatics. It was especially fun when cousin Ellen would come and show us stunts she did on her jungle gym at home.

- Down by the creek bottom, which we named Red Rock Canyon for some of the red soil where the creek made a bend, (and probably where the clay came from for the bricks for the house!) we made a make-shift clubhouse/tree house and furnished it with discarded junk.
- We also had a clubhouse up in the loft of the tool shed.
- In the winter we made snow forts, dug tunnels in the snow along the road banks, and looked for the tallest icicles we could find hanging from the barn roofs.
- Occasionally, Mom would treat us by bringing home a coloring book for us from the grocery store.
- There were always kittens or puppies to pet, and sometimes Mom would join us sitting on the steps by the back door.
- The boys tried their luck at rabbit and squirrel hunting and even frog gigging. None of us girls wanted to cook or eat them though!
- We even had a pet parakeet that Mom allowed to come out of its cage occasionally. When she would be ironing, it would fly to the ironing board and peck at its reflection in the iron. It would also like to fly to the chrome edged kitchen table, carry a ring in its beak to the edge of the table, and then drop it!

Esther in her Hecker dress sitting on the back door steps

David, Kathy, and Tom with the hog pen in the background, 1956

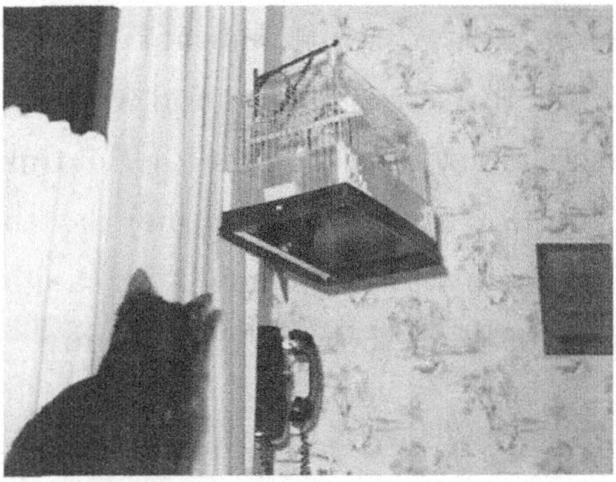

There were always dogs and cats on the farm, but normally they weren't allowed in the house as you see in this picture!

- We played "handy over" with a softball over the garage. The softball was thrown over the garage. If you caught it before it hit the ground, you could sneak around the garage and tag the other team to score a point.
- We also played baseball, basketball, and croquet.

Roger playing croquet, 1958

- On summer days we often swam in the pond and would dig in the 4 to 5-inch deep mud with our toes and come up with a clam. Dad even built a floating dock made from empty barrels. One summer day the swimming came to an abrupt end when Tom caught a baby catfish in his swim trunks! It was quite funny.

Kathy, Tom, Roger, Steve, David, 1957

Still using the raft in 1962!

Brother Steve submitted a list of memories growing up on the farm:

- Opening corn fields by hand with the horses and wagon, and later with a tractor and wagon and/or the pick-up truck.
- Early memories of big harvest days when folks would come out to help with use of a stationary threshing machine and bailer, not much recollection of details, but the wheat would first be cut, loaded on wagons, and then hauled in to run through the machines.
- Watching Mom and Dad milk the cows by hand before the mechanical milking machines were added.
- Going out to the woods and/or along fence rows to cut cedar Christmas trees the week or so before Christmas. Decorating the tree, going to the Christmas Eve service, and coming home to find that Santa had been there.
- The car trips with most of us kids in the back seat to visit Aunt Martha and Uncle Otto. Martha would always have fresh baked homemade bread; we would play with a big wheeled baby carriage; and we needed to use the outdoor toilet. It was fun listening to Uncle Otto talk about the towns he would frequent and working on the pipeline.
- Likewise, cramped car rides for day trips and picnics to the St. Louis Zoo, Fort De Charters and Fort Kaskaskia. Also, major day trips to the State Fair in Springfield.

- Riding with Dad to take pick-up loads of wheat to Waterloo and/or New Athens, and getting a cold bottle of coke for a nickel. Also, going to the county fair in Waterloo; it was near the dairy plant where we could get freshly made chocolate milk and ice cream.
- Fall apple butter making days with women helpers coming the night before to peel the apples.
- Hog butchering days with relatives (mostly with Eddie Guebert and Uncle Bank) coming to help and Dad (and some of us) going to their places to do the same. Eating some freshly boiled meats as the day ended, like kidneys.
- Breakfast with hot oatmeal and Webconbrot (blood sausage or klump).
- Getting some loads of coal into the basement for really cold nights, but mainly the making of firewood for heating the house. Dad would cut down trees, split logs, and we would have extra help to cut it into smaller pieces to fit into the furnace.
- Getting and raising baby chicks (filling the small feeders and waterers) into broilers and egg layers, gathering eggs, and getting the eggs ready for early morning weekly pickup.
- During summer months, our Sunday noon meals were usually delicious fried chicken. Judy helped Mom butcher three chickens every Saturday night. Mom would chop the heads off, and we'd dip them in boiling water to pluck the feathers before cleaning and cutting

them up. Then Sunday morning, after milking, Mom would fry the chicken and put it in a roaster while we were in church.

- Playing "cowboys" with hand carved guns in the woods and "red rock canyon" and building a log cabin.
- Playing in the snow: sledding, making snowmen and creating tunnels in drifts along the road banks.
- Attending relative wedding receptions where everyone seemed to be singing German beer songs—in particular, the Schnitzelbank song.
- Climbing up the ladder outside of the silo, watching others helping out in the making and blowing silage into the silo. Later we boys would go up and throw silage down the chute to feed the cows before going to school.
- Our Saturday night bath with "used water."
- Playing in the hay loft of the old barn; in later years we would sweat like pigs unloading and stacking hay under the hot tin roof of the new barn.
- Walking home a few times, the five miles from grade school in Red Bud.
- Running barefoot in the fresh furrows as Dad plowed the yard and nearby fields.
- The big fire and rebuilding into a modern dairy farm.
- Going "swimming" and fishing in the pond.

One summer, Dad plowed up the yard so that he could plant grass seed. Judy and John were following the plow barefooted, enjoying the cold, newly plowed furrow. In fact, we all went barefoot most of the summer. However, John stepped on a piece of glass in the furrow and cut his heel very badly. That put a stop to it! After warm spring showers, it was fun going barefoot down the muddy lane and all of us playing follow the leader with mud squishing between our toes.

If we had 3 or 4 days in a row of zero or below temps, Dad would check the pond to see if it was safe for ice skating. He would chop a hole and measure the ice; it was supposed to be at least four inches thick. This was shared ice skating as there were only a couple pairs of skates. Kathy remembers learning by putting on the skates and pushing a chair across the ice. One winter, Dad warned us that the ice was starting to melt, and not all of it was thick enough. Kathy, 4 or 5 years old, wanted to know which parts were thick enough, so she went around pointing and asking, "Is it frozen here? How about here?" She promptly found a spot that wasn't and fell through the ice! Judy pulled her out and carried her up to the house. Mom sat with Kathy in front of the open oven door to warm her.

We had activities and hobbies to keep us busy also.

- The girls, who belonged to the Rainbow Girls 4-H Club, took on numerous sewing, cooking, and craft projects. When Judy was old enough to join 4-H at 10 years old, her first project was a skirt and scarf made from a feed sack. She still has a tiny scrap of it in a little quilt she made for a doll.
- The boys each had heifers to take care of for 4-H projects and belonged to an agricultural 4-H club.
- John had sheep for an FFA project, and he took accordion lessons.
- Judy, Lois, and Kathy took piano lessons.

- Steve was very creative at designing and building structures out of toothpicks and balsa wood.
- David put together many model cars and trucks.
- Tom played basketball in grade school and also in high school.
- Roger added bandy chickens to the project list and played on a Hecker baseball team, getting himself there by riding the bicycle.

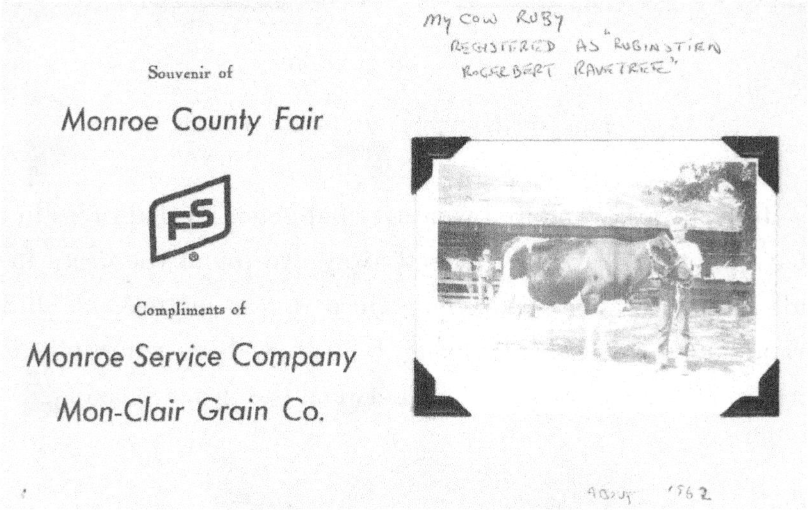

Roger at the Monroe County Fair

DNESDAY, AUGUST 8, 1956

Modeling at the state fair style show in Springfield will be Monroe County 4-H girls who were chosen for the honor last Wednesday at the annual Monroe County 4-H Show. The dresses were presented in a style show that evening.

Left to right are Louise Eisenhauer, Clara Ann Werth modeling Marlene Schaeffer's outfit, Evelyn Lemmers modeling Ramona Chattillion's pajama outfit, Ellen Kleyer, and Judy Gueber[illegible]. Miss Schaeffer and Miss Chattillon were unable to be present for the show. —Republican Photo

Judy (on the right) with 1956 4-H outfit

Judy always wondered whatever happened to that dress in the above photo. After Mom passed away, we found the dress in an old trunk in her bedroom. Here is a picture of Judy's granddaughter, Jessie, wearing that very dress some 63 years later in 2019! The resemblance between the two is amazing!

4-H Girls Chosen to Represent Monroe County At State Fair in Springfield Aug. 12-14

Nine girls will represent Monroe County at the State Fair Aug. 12 to 14. Those who will model in the 4-H Dress Revue are Mary Christopher, Sunflower Maids Club, Ella Louise Weitkamp, WHS Mix and Match Club, Ramona Chattillion, Lucky Clover Club, Linda Rehmer, WHS Mix and Match Club, and Lois Guebert, Rainbow Girls Club. Alternates for the Dress Revue were Marjorie Ann Wedel, Can Can Club, and Vicki Snodgrass, Stitch and Chatter Club. June Mueller of the Stitch and Chatter Club will give her demonstration on Cherry Custard Pie at the Fair. Girls who will attend classes in demonstrations will be Carol Hicks and Anna Mueller of the Stitch and Blaze Club and Penny Walker of the Can Can Club.

Exhibits sent from the county include a lunch cloth made by Ruth Ann Stumpf of the New Hanover Lucky Club, and garments made by Nancy Mueller, Can Can Club, Judy Guebert, Rainbow Girls Club, and Mary Kathryn Conrad, Sunflower Maids Club.

Mrs. Carl Guebert, leader of the Rainbow Girls 4-H Club will accompany the girls as chaperone. The girls will stay in the Junior Home Economics Building at the West end of the fairgrounds. Visitors to the fair should plan to visit this building to see the exhibits, and to watch the daily program in the auditorium. Monroe County girls will model in the public dress revue in the auditorium at 3 pm., Aug 14.

Models for the fair and clothing exhibits were chosen from an outstanding group of 25 girls. Others in this group were Karen Trost, Mary Kinzinger, Penny Walker, Phyllis Rehg, Theresa Steingruby, Ruth Ann Stumpf, June Mueller, Mary Ruth Pieper, Gertrude McCarthy, Jeanette M. Jehling, Annette Chattillion, Darlene Matzenbacher, Linda Lou Schmidt, Pat LaForce, and Jeanette Mueller.

Other girls who rated A's on their demonstrations and were awarded Banker's Ribbons at the County Show were Arlene Wetzel, Ella Louise Weitkamp, and Marjorie Ann Wedel.

Chosen at the annual county 4-H show July 31 to model their garments at the State Fair were Mary Christopher and Lois Guebert, seated, and Ella Louise Weitkamp, Linda Rehmer and Ramona Chattillion.

The clothes that these girls made will be sent to the fair for judging. Seated are Vicki Snodgrass, alternate, Mary Catherine Conrad and Phyllis Rehg; and standing, Nancy Mueller, Margie Wedel, alternate; and Judy Guebert.

Lois and Judy's 4-H outfits in 1958

STATE FAIR REPRESENTATIVES PICKED

Annual 4-H County Show Held At S.P.P.S. Gym Last Friday

Monroe County will also be represented at the fair by these girls. Left to right are Kathy Brandenberger, foods demonstration alternate; Kathy Guebert, food demonstrator, and whose garment will be judged; Jan Smith, whose garment will be judged; Kathy Friedrich, flower arrangement exhibitor; Susan Rick, dress revue alternate; and Cindy Rippelmeyer, dress revue alternate.
—Republican Photo

July 17 and 19 were days of activity for 4-H'ers enrolled in Home Economics projects, as they participated in the 4-H County Show held in the SS. Peter and Paul Grade School Gymnasium in Waterloo.

Wednesday, July 17, the clothing projects were judged. Twenty four girls were rated superior as they modeled their garments before the judges in the State Fair Elimination Contest. They were Kathy Stiening, Linda Koch, Judith Hesse, Brenda Mehrtens, Barbara Mehrtens, Eileen Offermann, Patricia Sensel, Sheryl Sensel, Cindy Rippelmeyer, Kay Rippelmeyer, Marla Rippelmeyer, Linda Roever, Joan Schulden, Jan Smith, Donna Heller, Nancy Mehrtens, Susan Telford, Harriett Zipfel, Betty Braun, Kathy Guebert, Rita Kaiser, Sandy Rippelmeyer, Judy Schewe, and Susan Rick.

Four-H members enrolled in Foods, Room Improvement, Photography, Baby Sitting, Arts and Crafts, Party Planning, Indoor Gardening, Flower Gardening, Pebble Pups, Career Explorations, Dog Care, Bird Study, and Flower Arrangement exhibited representative samples of their project work on Friday, July 19.

On Friday evening the girls modeled their garments in the Dress Revue. Junior leaders who served as narrators were Judith Hesse, Rita Kaiser, Glenda Esker, Sheryl Sensel, and Jan Smith.

Paul Kennedy, representing the St. Clair-Monroe Bankers' Federation, presented Bankers' Ribbons to Kathy Guebert and Kathy Brandenburger. These girls were rated A plus in the Food Demonstration Contest held July 5.

Announcement of the State Fair winners was made by Mrs. Al Hoffmann, Extension Adviser in Home Economics. Representing Monroe County at the State Fair in August will be: In Dress Revue, Rita Kaiser of the Rainbow Girls 4-H Club, Harriett Zipfel of the Po-ettes, Glenda Esker of the Friendly Neighbors, Kathy Stiening of the Blue Bonnets, Eileen Offermann of the Friendly Neighbors, and Marla Rippelmeyer of the G. G. 4-H Club. Alternates are Susan Rick of the Warnock 4-H Club, and Cindy Rippelmeyer of the G. G. Club. Kathy Guebert and Jan Smith will send their garments to the State Fair to be judged on construction. The Food Demonstrator will be Kathy Guebert. Kathy Brandenburger will attend Foods Classes as the alternate in Foods Demonstration. In Flower Arrangement, the exhibitor will be Kathy Friedrich.

Kathy's 4-H outfit in 1968

Dad often sat in his easy chair in the dining room reading in the evening after milking. Unfortunately, the rest of us never did much recreational reading growing up. We got the daily newspaper (the Globe Democrat) in the mail, but we had very few books at home, and obviously Mom had no time to read to us at bedtime! The family radio was set for KMOX, even in the milk barn for the cows! We never had a record player until John brought one home for Christmas after he joined the Navy in 1960.

The first family telephone was a big hand cranked one attached to the wall on the left of the bay window. It was a party line which meant other neighbors on the line could listen in on our conversation! The number and length of rings you heard determined whether the call was for the Gueberts, (long, short, long for us) or another neighbor. Even after it was no longer a party line, the single family phone was for important matters only. You never used the phone just to chat with someone. If you called anyone outside the immediate Hecker area, it was a toll—10 cents to call Red Bud. So, you had better have a very good reason before you made that call!

Growing up on the farm included some accidents. Once when Judy was playing acrobats on the floor with 5-year-old David, she flipped him off her feet and he flew across the room and hit the heater vent. His right arm broke and had to be put in a cast. He remembers stacking bricks to construct a make-believe milk house a couple days later. Someone tattled to Mom, and she promptly made him come in the house and said he couldn't be doing that with his arm in a cast. Later on, when David was in 7th grade, he was helping Dad split logs for firewood for the furnace. The head of the sledgehammer that Dad was using flew off and hit David in the forehead! Thankfully, the injury was not nearly as bad as it could have been. It knocked David out, and the cut required stitches, but there was no permanent damage.

Another time all the kids, except for Kathy, were waiting for Mom to pick them up after school. Roger was in 1st grade, Judy in 8th, all the rest in-between. Mom was running late, so they were all playing on the playground. Roger slipped playing tag and ran right into the path of the swing that Steve was on. The corner of the wooden swing seat hit him on the forehead, so of course there was blood. Big sister Judy carried him to Doc Mattingly's office about 8 blocks away for stitches!

Still another time, while Mom and Dad were both gone, brother Tom gave Judy a scare. We had an old horse named

Smoky that was not too fond of being ridden. Roger had already ridden him, and Tom decided to do so, also. Smoky had other ideas. As Tom was walking around the horse, Smoky kicked him hard in the chest and knocked him unconscious. Roger came running in to the house telling Judy, "Tom's dead. He's a goner!" Tom had the wind knocked out of him and had a couple broken ribs, but fortunately he was okay.

Roger riding Smoky in 1961

The kids were not the only ones to get hurt. Sometime in the early 1960s while using a chain saw to cut up logs, Dad's hand slipped, and he cut his knee severely, requiring many stitches. He was laid up for quite a while, requiring his sons to assume the workload during his recuperation.

One momentous, awful day occurred in August of 1955. Judy and Mom were school clothes shopping in Belleville, Illinois, when another shopper recognized Esther and said, "Aren't you Mrs. Carl Guebert?" Your barn is on fire!" Judy and Mom headed home, smelling smoke in the air as they got closer to Hecker, and realized it must be true. The big hay barn was burning, started by spontaneous combustion. If hay is put away damp, a chemical reaction can occur that builds heat, causing it to burn. Evidently, that is what happened. There was another barn fire the same day in Monroe County. Nowadays an acid is often sprayed on the hay to prevent this. Many people arrived, and fire trucks pumped water from the pond, but the barn burned down completely. Hired hand Henry helped get the work horses (Dave and Dolly) out of the barn. They had hot dripping tar on them, but they were okay. At first, Henry thought he was on fire and started frantically jumping up and down as the hot asphalt from the side of the chicken house had dripped down his back. Fortunately, he was okay. No one else was injured, and no cattle were lost. The fire was actually a blessing in disguise as Dad was able to build a modern milking parlor, a barn to house the hay, a second silo, and a feed trough.

Red Bud Pi

Friday, Sept. 2, 1955

Entered As Second Class Mail In The Post Office At Red Bud, Illinois

FIRE DESTROYS BARN—A fierce blaze, whipped by a strong wind, completely destroyed a barn on the Carl Guebert farm near Hecker Monday afternoon. The barn was filled with hay. Both the Hecker and Red Bud fire companies were called but the blaze had too much of a start. Efforts to save near-by buildings succeeded and the fire was contained in the barn. Guebert told authorities that he delieved spontaneous combustion was the cause of the fire.

Waterloo Paper

Gale Fanned Fires Destroy Guebert, Goeddel Dairy Barns Within Six Hours in Identical Fashion Monday

Under almost identical circumstances two prominent Monroe County dairy farmers' barns were burned to the ground Monday in fires that broke out about six hours apart.

About 3 pm. fire broke out at the Carl Guebert farm about two miles south of Hecker completely destroying the barn and milking parlor, and about 9:45 a strikingly similar fire broke out at the Harold Goeddel farm, about three miles east of Waterloo, burning the barn to the ground.

No livestock was lost at either place.

The start of each fire was described in an almost identical manner. A small blaze was seen, and then, almost immediately, the entire structure was in flames.

Both the Hecker and Red Bud fire departments were called, but the flames had gained too much headway for them to be able to do anything except protect nearby buildings. A nearby hay shed was in danger, but with plenty of water available from a nearby pond, the firemen were able to save it.

The blaze was fanned by a high wind that blew the flames away from the home.

There were about 50 men fighting the blaze. No one was injured, although one man suffered a burned arm when he got against a steel grain bin that was near the fire.

After the large pile of hay and grain had burned down somewhat Armin Reinhardt of New Athens used his bulldozer to push the debris apart and out into a field.

Hecker firemen remained on the scene until after midnight.

No immediate estimate was made on the loss, which was partially covered by insurance. Included in the loss were a $1700 milk cooler which was fully covered by insurance, milking machine, other milking equipment, 1000 bales of hay, 600 bushels of oats, and 100 bushels or barley.

The 35 ft. silo remained standing.

The barn had just been repaired and painted.

The cause of the fire is undetermined, but it is believed that it could have been spontaneous combustion.

The debris was still smouldering when work began on a temporary structure in which the herd of 20 Holstein cows could be milked.

The fire at the Goeddel farm broke out about 9:45 pm. Goeddel was outside sharpening chopper knives when flames broke out in one part of the barn. He said at first it didn't look too bad. He called to his wife to call the fire department and then went to get a garden hose, but with that, the entire structure burst into flames.

The Goeddel's telephone was out of order, and Mrs. Goeddel had to drive to Lawrence Williams place about two miles away to turn in the alarm since all the neighbors were on the same party line, and the entire line was out.

Goeddel said that when he realized the blaze was out of control he chased the few cows that were in the barn out and led out a small calf. He also drove a tractor out of the burning building.

Gusts of wind more than 40 miles per hour fanned the raging flames which ate quickly through the wooden 36 by 75 ft. structure, and when the Waterloo fire department arrived there was little to do except prevent the fire from spreading to nearby buildings.

The milking parlor was downwind from the burning barn and was being constantly showered with sparks. Firemen ignored the sparks and advanced against the intense heat to play water on the nearby buildings. The milking parlor had an asbestos roof and walls of concrete blocks. Also saved were two garages and a row of sheds. The heat was so intense that the paint on all the buildings was badly blistered.

A double concrete silo adjoining the barn also remained standing.

Goeddel said his loss was partially covered by insurance.

He estimated the value of the

(Continued on page 8)

FIRE

barn at $5000. In it, and destroyed were about 80 tons of hay and 800 bales of straw. He had no grain in the building and believes he can salvage most of the silage. It burned somewhat on the top, but the firemen scaled the silo and put that blaze out.

Fire chief Louis Nobbe praised the work of his men in controling the fire. They stayed on the job until about 3 am and then along with the Valmeyer Fire Department, and hooked their hoses and pumps together to pump from a distant pond to put out the still burning hay.

The burning building turned the low hanging clouds a fiery crimson which was easily visible for many miles. Men working at the Guebert fire could see the glow.

Cause of the Goedel fire, too, was a mystery with theories of spontaneous combustion and dust gathering on a light globe being expresed.

There were no injuries in fighting the fire, although some firemen suffered scorched faces, and one's hair was burned.

Beginning construction of the new feedlot

Completed feed trough

The fall of 1955 was momentous in another way. Judy started high school, riding the bus to Waterloo. Switching from the Lutheran grade school in Red Bud to the public high school in Waterloo was not an easy transition. She knew Willa Mae Wiegand Prange from her days at the one room schoolhouse through third grade, and then Eileen Probst Miller from 4-H, but no one else. Judy felt the high school kids were very cliquish. The transition for the rest of us was not easy either, but not as difficult because by this time, we had a little more interaction with other Waterloo kids through our 4-H involvement.

Judy joined the high school home economics club, called "Future Homemakers of America", and was active in it. She even got little sister Kathy and brother Roger involved, serving as attendants for the FHA Sweetheart Dance. Judy had fun sewing dresses for Kathy as home projects. In addition, Judy was active with Walther League, the high school youth group at church. The rest of us followed her lead later on, with Lois and Kathy joining FHA and the boys belonging to Future Farmers of America and all of us being active in the Walther League.

Kathy and Roger serving as attendants for the FHA dance

The 1950s brought other big changes. The first television we saw was at Uncle Honey's house (a black and white one) in 1953. The whole family went to Uncle Honey and Aunt Esther's in Red Bud to watch Queen Elizabeth's coronation. Dad finally broke down shortly afterwards and bought a TV from Mom's cousin Ernie Werner, who was a television repairman. We didn't get a colored TV until 1968. One nice feature back then - television watching was free! No cable company or satellite dish was needed. There were only about 5 channels that signed off at midnight and back on again at 5 am.

Mom enjoyed watching soap operas while she ironed. Back then, much of the clothing needed to be ironed! (Judy liked ironing the puff sleeves on Kathy's little dresses, pressing them out like a fan.) Favorite TV shows were Lassie, I Love Lucy, This Is Your Life, Lawrence Welk, Roy Rogers and other westerns, I've Got a Secret, Bonanza, and Walt Disney. Queen for a Day was fun to watch, but we all thought Mom should have been on that show, as she certainly deserved to be one of the "queens for a day." Going to a movie theatre was unheard of for us, with two exceptions. We went with a group from church to see "Seven Wonders of the World" and "The Ten Commandments" at a huge theatre in downtown St. Louis.

Vacations were nonexistent. Remember, there were cows to milk in the morning and the evening! A big day was going to the zoo or to a Cardinal baseball game, leaving after morning milking and being home before 4:30 or 5 for evening milking. Mom would bring fried chicken along for our lunch. Judy got to go along with Dad and Hank Burgdorf one Sunday afternoon to a Cardinals ballgame, when it was at Sportsman's Park on Spring Avenue. She also remembers going along to a Hawks game and to see the

Harlem Globetrotters. Dad allowed us kids to go to summer camp for a week if we wanted to go. Most of us took advantage of this, be it church camp or 4-H camp. Judy, especially, lived for that week in August at the Lutheran school camp at Camp Piasa, Pere Marquette State Park. (It was a vacation—away from work and the rest of the family!)

Another big day was going to visit relatives or their coming to visit us, usually on a Sunday afternoon. Remember, we never rode in a car for very long. Who would want to do so with 10 in a car! Kathy usually sat on Mom's lap, and one or two more sat between Mom and Dad with the rest crammed in the back. To fit, we sat staggered, two sitting on just the front part of the seat, and three sitting all the way back. Dad never took our suggestion to buy a station wagon even though we pleaded and begged for one! There were no such things as car seats or seat belts back then. We were wedged in tight without them!

Maybe that is why Tom got left at church one day. No one missed him in a crowded car! It didn't bother six-year-old Tom. He knew his way home (5 miles away) so he just started walking. Roger Brandt and his wife picked him up along Route 159 about a mile from Red Bud and brought him home, arriving home before the rest of the family with just hired hand Henry at the house. Tom remembers the couple asking Henry if Tom was at the right place. Henry looked at him and said, "Yes, that's Tommy!" Dave also got left behind once, as we were returning from a visit to Uncle Walter and Aunt Elsie's in Millstadt.

It used to be that Confirmation day was the day a girl felt she entered womanhood. She got to wear seamed or seamless stockings with a garter-belt. Panty hose weren't popular yet. You also wore white gloves and a hat to church, especially on communion day. Judy was confirmed in 1955 and remembers communion was served once every other month. You had to announce your intention to commune sometime the week before. She remembers going into town one night with Dad after milking to Pastor Gross's house to let him know we wanted to commune the next day. These customs went by the wayside by the early '60s. We no longer wore hats and gloves to church, communion was offered monthly, and one did not have to announce ahead to commune.

All 8 children at a family gathering at Uncle Vic and Aunt Ruth's, 1955

Dad and Guebert Uncles: Honey, Vic, Bank, Dick, Walks, 1955

Back row: Aunt Alma, Mom, Big Aunt Norma, Aunt Ruth
On bench: Aunt Esther, Cousin Lorraine, 1955

As we get ready to enter another decade, another milestone is reached. The first child graduates from high school! Judy graduated in 1959.

Judy receiving her H.S. diploma from Dad, School Board President

At that time, not many went on for further education, unless you wanted to be a teacher or nurse. A secretarial job at the American Furnace Factory (AFCO) in Red Bud was available at the time for Judy. Cousin Rosie Ehlers was quitting as she was expecting a baby. So, Judy started at AFCO and was able to catch a ride at the end of the lane with other AFCO workers. She kept living at home and didn't need a car. Judy said later she wished she had experienced living on her own a while before getting married. However, Kathy was delighted to have her big sister still at home and would walk to the end of the lane to meet her when she came home from work (aww—how sweet).

CHAPTER 4

THE 1960s

The 1960s brought many events, including seven high school graduations, three college graduations, two weddings, and two grandchildren!

With the boys older and able to handle milking on their own, Dad had a bit more freedom and accepted a position with the Illinois Agricultural Association. This involved some overnight stays in Bloomington, Illinois, and an annual convention in Chicago, which Mom also attended.

With the children being older, Esther also had more freedom and got more involved as a 4-H leader. Besides hosting the monthly 4-H meetings at our house, she arranged field trips and set up sewing clinics (with as many as 4 sewing machines going at a time!) in the dining room. Many of her students, including us three girls, earned overnight trips to the Illinois State Fair in Springfield, both with sewing projects and food demonstrations. She had a positive impact on her 4-H girls, as evidenced by the many who came years later to her visitation when she passed away. Mom's mind was often working in the middle of the night, coming up with ideas for "Share-the Fun" skits when she should have been sleeping! Some of those skits also earned us a slot at the Illinois State Fair. Mom was an accomplished seamstress herself and showed her expertise with the "Make It Yourself with Wool" contests.

Carl Guebert: Farm Bureau leader, school board president, and dairy farmer.

Meet Your New IAA Director

CARL E. GUEBERT

"One of the most important tasks facing small town school systems is preparing rural youngsters for off-the-farm opportunities."

That is the opinion of Carl E. Guebert, Monroe County dairy farmer and Illinois Agricultural Association director. He is president of the Waterloo Community Unit School District.

"High schools should provide more than vocational agriculture training for boys. The schools should introduce youngsters to various crafts and trades, since many of the students will be employed away from the farm. I feel that counseling and guidance programs to help youngsters prepare for careers should be expanded."

Guebert has served on the Waterloo school board for nine years and has been its president for the past four. The unit district includes schools at Hecker, Renault, and Waterloo, which are attended by 1,300 students.

Guebert and his wife, Esther, are especially interested in education and activities of children because they have eight of their own. Kathryn, the youngest, is a sixth grader. Roger, Thomas, David, and Steven are students at the Waterloo High School. Lois is a freshman at Southern Illinois University, Carbondale. John is in the Navy. Judith, the oldest daughter, is a secretary in Red Bud.

The high school lads give their father considerable assistance on his 200-acre farm, two miles south of Hecker. The farm is primarily a dairy operation, although crops include corn, alfalfa, wheat, and soybeans.

Guebert started dairy farming in 1939. For the past 16 years he has received the Purebred Dairy Cattle Association national honor roll award in recognition of the achievement of his dairy herd. His dairy herd improvement records show that the 36 Holsteins in his milking herd produced an average of 517 pounds of butterfat and 13,000 pounds of milk last year.

A feature of his dairy operation is an automatic electric mix mill which grinds grain at certain intervals and then blows the grain through tubes to the milking parlor. The system saves time and money.

Active in Farm Bureau for more than 22 years, Guebert served six years on the Monroe County Farm Bureau board and six years on the board of the Monroe Service Company. He was Farm Bureau secretary for five years and president of the service company for four years.

Guebert feels that Farm Bureau has an important role to play in the promotion and marketing of farm commodities. "We can produce the stuff, now we must sell it."

Carl Guebert, president of the Waterloo Community Unit School District Five, and W. J. Zahnow, school unit superintendent.

"Counseling and guidance programs to help youngsters prepare for careers should be expanded."

While visiting Waterloo High School, Guebert chats with his sons, from left, Thomas, Steven, David, and Roger.

He was elected district 16 director of the IAA in December to succeed Harold E. Hartley, Centralia, who became IAA vice president.

Guebert is secretary of the Monroe, St. Clair, Randolph Dairy Herd Improvement Association and is a member of the American Dairy Association and the Southern Illinois Breeding Association.

He was an original director of the soil conservation district in Monroe County and has served for four years as chairman of the county's agricultural extension council.

Guebert is a member of a five-man county zoning committee which was organized more than a year ago to promote interest in zoning Monroe County. The group has worked with the county board of commissioners and others to stimulate interest in the proposal.

Guebert feels that zoning is a form of protection rather than control. "Zoning, an orderly way of planning, protects property values."

The Gueberts are active in the St. John's Lutheran Church of Red Bud. Carl is chairman of the congregation.

Although farming, school board meetings, and organizational activities take up the bulk of Guebert's time, he is an avid sports fan — a rooter for the teams at nearby St. Louis, Missouri. During the winter he follows the St. Louis Hawks basketball team, and during the baseball season he favors the Cardinals.

Mrs. Guebert is also active in a number of organizations. She has been a member of the Home Bureau board for three years and has served as its secretary for two years. She has held the positions of president, vice president, and secretary of the Monroe County Homemakers Extension Council. She is a 4-H club leader and a former Sunday School teacher.

What is down the road for agriculture? Guebert is convinced that fewer men will be farming in the future, but he is optimistic about the future of the family farm.

"A two-man operation is still the most efficient farm. Family farms are assured a place in the future for that reason." □

In addition to caring for her family, Mrs. Guebert participates in Home Bureau, 4-H, and church activities.

25

On the next page is a newspaper clipping from one of the wool contests in 1966, which both Mom and Kathy entered.

REPUBLICAN, WATERLOO, ILLINOIS, WEDNESDAY, NOVEMBER 2, 1966

Winners from Monroe County in the District "Make It With Wool" contest, Oct. 22, at the Farm Bureau Building in Belleville were, left to right, Kathy Stumpf, Brenda Mehrtens, Kathy Guebert, Mrs. Carl Guebert, and Mrs. Charles Hesse.

Local Winners In District 'Make It With Wool' Contest

Five Monroe County women participated in the District "Make It With Wool" Contest held Oct. 22 at the Farm Bureau Building in Belleville.

Mrs. Helen Heise, Waterloo, placed first, and Mrs. Carl Guebert, Hecker, second in the Adult Contest of 15 entries.

Kathy Stumpf, daughter of Edmund E. Stumpf, Columbia, placed fifth in the Senior Class of 13 entries.

Kathy Guebert, daughter of Mr. and Mrs. Carl Guebert, placed first, and Brenda Mehrtens, daughter of Mr. and Mrs. Merle Mehrtens, Fults, placed ninth in the Junior Class of 19 entries.

Miss Patricia Mallet, Randolph County Home Adviser, Frances Axley, Waterloo High School Home Economics Instructor and Kay Vancil, Washington County Home Adviser were judges.

Mrs. Hesse and Kathy Guebert will represent the District in State Competition to be held at the Sherman Hotel in Chicago Nov. 15.

The Contest is sponsored by the Illinois Farm Bureau, The National Wool Growers Association and the Illinois Purebred Sheep Breeders Association, and is designed to encourage the use of wool.

John graduated in 1960 and enlisted in the Navy. He served for three years and traveled all over the world on a tanker ship. When he was discharged in 1963, he lived at home for a couple years, working for an experimental research farm and then in March 1964 for the National Cash Register Company in St. Louis, where he later got an apartment.

Photo of all 8 of us when John was home on leave

While in high school, Lois was very active in FHA, serving as a state officer her last year of high school. She graduated in 1962 and enrolled at Southern Illinois University in Carbondale, majoring in home economics education. She graduated from college in 1966 and took a teaching position at R.O.V.A. High School in Knoxville, Illinois.

While Judy was working at the furnace factory, cousin Dorcas (Aunt Florence and Uncle Hans Mueller's daughter) set her up on a blind date with Charles Meyr. (Charles was also Dorcas's cousin, but on her dad's side! Hans was a brother to Charles's mother, Annette Mueller Meyr.) Charles and Judy hit it off, and on May 11, 1963, they were married at St. John Lutheran Church in Red Bud with a reception out at the farm.

Judy and Charles' wedding, May 11, 1963

In the spring of 1963, Dad quit raising chickens. He knocked out 2 walls of the chicken house and we white-washed the walls. The building, having a concrete floor, made an excellent hall for

Judy's wedding reception. Many guests thought this reception in the old chicken house was so fun and most memorable. Kathy thought this was so cool that she decided right then if she ever got married, she wanted her wedding reception on the farm too. Hers was held on the huge front lawn and the new front porch, also very memorable.

Steve graduated in 1963 and enrolled at the University of Illinois in Champaign-Urbana, majoring in agricultural economics. He graduated from the U of I in 1967 and chose to work on his master's degree and then his doctorate at the University of Wisconsin in Madison. At one point we thought he was going to be a perpetual student! He did very well academically, earned some scholarships, and graduated with very little debt. (College costs were much lower back then!)

David graduated from high school in 1964 and enlisted in the Army Reserves that same Spring. He had six months of training at Fort Knox in Louisville, Kentucky. Then he was back on the farm working with Dad except for short tours of duty for the next six years.

August 14, 1964, Carl and Esther's first grandchild, Jodi Lynn Meyr, was born to Judy and Charles. It was great fun for everyone having a baby in the family once again. Two years later, grandson Randall Charles (Randy) Meyr was born on June 28, 1966. Judy and Charles' second son, Jeremy James, was born May 14, 1981.

Tom graduated in 1965 and enrolled at the University of Illinois in Champaign-Urbana, majoring in math education. He had an Illinois Teacher's Scholarship and also worked on campus so was able to graduate in 1969 with no debt.

Carl and Esther with Judy and first grandchild, Jodi, preparing for 25th anniversary celebration dinner.

Roger graduated in 1966 and before he even finished high school, he enlisted in the Marines, much to the dismay of Dad. The Vietnam War was going on, so Dad was undoubtedly worried about that. Dad had never had to serve in the military. World War II was going on when he and Mom were first married, but he was not drafted because of his occupation as a farmer. John and David were both in the military but never had to serve in combat.

In Vietnam, Roger went through some very tough times which he shared in letters to Judy, so as not to worry Mom. He did survive, but he was seriously injured and earned two purple hearts. Kathy remembers the day a uniformed Marine officer knocked on the front door to inform us that Roger had been injured. That was also how they notified family if someone had died, so there were some very scary moments until we learned of the extent of his injuries.

Family picture with Roger home on leave in 1967

Roger Guebert Hurt in Action

Marine Lance Cpl. Roger Guebert, son of Mr. and Mrs. Carl Guebert of near Hecker, was wounded in action New Year's Day near Quang Tri, Vietnam, according to word received here by his parents.

Mr. and Mrs. Guebert were notified the following day by a Marine non-commissioned officer, and Thursday they received an official telegram.

According to the telegram Cpl. Guebert suffered wounds to the right leg, abdomen, and right eyebrow inflicted by a "hostile explosive device," while he was on patrol.

He is being treated at the Third Medical Batallion Hospital.

Cpl. Guebert entered the Marines in September of 1966 and has been in Vietnam for six months.

Newspaper article about Roger

Kathy graduated in 1969 and followed in Lois's footsteps, enrolling at SIU, Carbondale, in home economics education. Now it's called Family and Consumer Science! She, too, had an Illinois Teacher's Scholarship, worked part time throughout college, and was able to graduate with no money, but also with no debt! True, as we said earlier, college costs were much lower then, but we also made do with a lot less. For example, Kathy lived in a one bedroom, one bathroom apartment with three other girls, two bunk beds, one TV, one house phone, no computers, and no cell phones!

The second family wedding was August 16, 1969. Tom married his high school sweetheart, Barbara (Cookie) Osterhage. He then took a math teaching position in Mount Vernon, Illinois.

Tom and Cookie's wedding, August 16, 1969

CHAPTER 5

THE 1970s

The 1970s brought many changes also. There were no more high school graduations, but there was one more college graduation, five weddings, 10 more grandchildren, and the creation of the Escarda Dairy!

In 1970, David's tour of duty with the Army Reserves was over, so he told Dad he either wanted to form a partnership or look for work elsewhere. Dad said if David would give him $50,000, they would have a deal. David went to First State Bank and talked to Adrian Rathert about getting a loan. At first Adrian said David could get the loan if Carl would cosign for it. However, the next day he called back and said the bank could not do it because Carl was President of the Bank Board. David then went to the First National Bank where Leonard Vogt was the loan officer. Within a few minutes David had a check for $50,000! Leonard said cosigning was not necessary; if David wasn't good enough for the money, then no one was. That was the beginning of Escarda Dairy, named after Esther, Carl, and David. Sister Lois by the way, is the one who came up with the name.

During the 1970s, Esther switched from being a 4-H leader to volunteering with the Red Bud hospital auxiliary and continued with this until the year 2000. She helped manage the gift shop and also made many crafts for the hospital bazaar.

While working for NCR, John met his future wife. He married Judy Fish Stuckstedde on September 3, 1971, and became an instant father to Judy's son, Mark Edward, who was born April 8, 1969, and whom he later adopted. Twin girls, Kimberly Ann (Kim) and Kelly Renee were born December 21, 1972. Son Gregory Scott (Greg) was born November 15, 1975, and son Christopher Paul (Chris) was born May 18, 1979. Daughter Jaime Linn was born July 6, 1982.

John and Judy's wedding, 1971

Late summer seemed to be the time for weddings. A year after John's wedding, on September 1, 1972, Roger married Joyce Posey, whom he met at a dance in Waterloo. Roger continued working for Golden Dipt Breading in Millstadt, Illinois, (where Mom lived as a teenager!). Daughter Jaclyn Marie (Jackie) was born February 14, 1977. Son Scott Edward was born April 28, 1981. Roger and Joyce divorced in 2000.

Roger and Joyce's wedding, 1972

Lois had been bridesmaid in countless weddings and decided she had had enough of all the "hubbub." So, when she and Joel McQueen, whom she met through a friend, decided to get married, they eloped to Las Vegas and were married June 10, 1973. They opted out of a big formal wedding, but Dad and Mom still hosted a reception for them on the farm.

Joel and Lois' reception, 1973

Later that same summer, Dad and Mom hosted another wedding reception on August 18, 1973, as Kathy married Thomas Murray (Tom), whom she met while in college in Carbondale. (Tom likes to tell the story about picking her up in a bar, but we won't go there!) Their first home was an apartment in St. Anne, Illinois, where Tom worked as a farmhand and Kathy taught home economics at the high school. Interestingly, after 10 moves and 27 years, they ended up in St. Ann, Missouri! James Carl (J.C.) was born May 12, 1978. Daughter Anne Marie (Annie) was born May 13, 1981, just one day before cousin Jeremy!

Kathy and Tom's wedding, 1973

Wedding number five for this decade was August 24, 1974, when David married Alice Driver. Roger and Joyce introduced the two of them via motorcycles! David and Alice built a house on the farm. Alice first commuted to work in St. Louis but soon was completely enmeshed in the farming operation, working side by side with David. She started milking full time with Dave when Dad retired on his 65th birthday, January 4, 1979. She was a city gal who originally said, "I will never like a farm and milk!" Daughter Becky Jean was born July 18, 1978. Two more daughters followed: Traci Anna on December 30, 1980, and Karla Lynn on September 30, 1984.

David and Alice's wedding, 1974

This is a picture of the Werner aunts and uncles taken in 1974
at Dave and Alice's wedding.
Front row: Aunt Tina, Aunt Florence, Aunt Elsie, Aunt Martha
Back row: Uncle Ray, Aunt Vicki, Uncle Hans, Dad, Mom, Uncle Walter, Uncle Otto

This is a picture of the Guebert aunts and uncles taken in 1974
at Dave and Alice's wedding.
Front row: Aunt Esther, Aunt Ruth, Mom, Aunt Norma, Aunt Alma
Back row: Uncle Honey, Aunt Edna, Uncle Bank, Dad

David said that a lot of people think or know that it is hard to work with your spouse, but he can tell you it is very hard to work with your father as a partner. In December of 1978, Dad, who would be turning 65 in two weeks, told David and Alice to come over to the house after milking because Leonard Vogt would be there to talk about the farm. David had just assumed that he would rent the farm when Dad retired. Dad had other plans! David and Alice were shocked when Dad said they would have to buy the farm if they wanted to continue milking cows and that he wasn't going to milk one more cow after his birthday! Mr. Vogt told them to each write down what they thought the farm was worth. Carl wrote $1,600 per acre, Esther $1,500, and David $1,800. Mr. Vogt said, "David bought the farm!"

CHAPTER 6

The 1980s and Beyond

Dad retired in 1979, and brother David took over full control of the farming operation. Dad and Mom continued to live in the farmhouse, and Mom babysit for David and Alice's girls during the morning milking. Now Dad and Mom were free to travel, and travel they did. They took trips to Germany, Hawaii, and Alaska.

Dad and Mom now had more grandchildren and even great-grandchildren to enjoy! To make it a little easier to keep track of who belongs to whom, we are going to group by sibling families in the "Family Trees" chapter, number 11.

Retirement life was not without some health issues, though. In 1983 the headaches Mom had been having for a number of years got considerably worse, prompting a visit to a neurologist. It turned out she had a slow growing tumor pressing on her brain. She was referred to a doctor in Memphis, Tennessee, who had just learned how to do laser surgery. So, they headed to Memphis, and on November 19th she underwent surgery to remove the tumor. It wasn't without concern, as she suffered a small stroke while in recovery. She was in the hospital over Thanksgiving that year. Those of us who were there will remember a Thanksgiving eve service that we attended with gratitude that Mom had survived the ordeal. The doctor explained there were several "fingers" fanning out from the mass that he could not remove. So, there was potential that the tumor would grow back. She lost partial vision in her right eye because of nerve damage, but the headaches were gone, and she was still able to drive. She even took up "Bob Ross" painting after her recovery, and all of us kids have at least one of her paintings.

This is Judy's recollection. "I'll always remember one Thanksgiving gathering in the late '80s. After a delicious turkey dinner, Mom got up from the table and announced that she had painted and framed a bunch of her paintings. She had them hanging throughout the house—upstairs and down. They were all numbered, and we were to vote secretly which painting we preferred. If more wanted the same one, she would try to duplicate it, and we'd receive it at a later date. Sure enough, about a month later, Mom was at the door with a painting for me. Each one of us received a treasure."

In 1985 Dad had a heart attack and underwent triple by-pass surgery at Barnes Hospital in St. Louis. He recovered well and continued to lead an active life for five more years.

There was one more wedding. On April 7, 1989, Steve married Carol Jenson in a small ceremony in Arlington, Virginia. Mom and Dad flew out for the wedding and later, on June 17th, they hosted a reception for them on the farm. Steve became an "instant father" as Carol had two sons from a previous marriage, David and Phillip.

In February of 1990, just 3 months short of their 50th anniversary, Carl suffered a stroke while he and Esther were bowling with their weekly senior citizens group in Red Bud. He never regained consciousness and died on February 23rd at the age of 76. The line for viewing at his visitation was extremely long, and he was eulogized as a wonderful servant of the Lord.

Steve and Carol's wedding reception on the farm, 1989

Esther continued to live in the farmhouse for 17 more years and was very active until the last few years. She visited Norma (Dad's sister who was in assisted living in Red Bud) almost daily until Norma passed away in 1994. She kept herself busy with church activities, as well as working in the St. Clement's hospital gift shop and auxiliary.

In 1998 Esther started having trouble swallowing as the tumor had grown down into her throat. She underwent surgery once again by the same doctor in Memphis. The surgery was successful, but after two years, nerve damage from it left her unable to swallow, requiring a feeding tube. She was not able to swallow even a drop of water or eat anything for seven years, but she never complained! In fact, when Pastor Seger would come to visit her, he said she lifted his spirits! She had the same gracious spirit, humbleness, contentment, love, and strong faith that her mother Katie displayed many years before when confined to a wheel chair.

A highlight for Mom in her later years was a Werner reunion at the farm on Labor Day in 2002. Her brother Walter, cousins Bobby, Herbie, and Ernie Werner, nieces, nephews, and children attended from as far away as Georgia, Oregon, Texas, Vermont, Ohio, Missouri, and Illinois. After saying grace before dinner while it was still very quiet, Mom, in her gravelly voice thanked everyone for coming that day and not waiting for her funeral. That also happened to be the day the Mueller cousins told us about the "Hecker Dress."

In 2007 Esther was diagnosed with congestive heart failure. She spent three weeks in the Waterloo nursing home before she died on May 31, 2007, at the age of 89.

John, Roger, David, Kathy, Lois, Judy, Steve, Tom after Mom's funeral, 2007

Carl and Esther are now both in heaven and no longer with us here on earth. However, their legacy continues with their children, grandchildren, and great-grandchildren.

Carl and Esther's descendants continue to gather for an annual Guebert Family Christmas, usually the Saturday after Christmas at St. John's Lutheran School in Red Bud. Our family has grown to over 80 members, and we usually have about 60 attending the celebration.

Carl and Esther's children and spouses present at Christmas 2012 celebration (Wearing "Got Milk?" headbands)

Some of Carl and Esther's grandchildren present at Christmas 2012 celebration

Carl and Esther's children and spouses present at Christmas 2018 celebration

Carl and Esther's grandchildren and spouses present at Christmas 2019 celebration

Carl and Esther's great grandchildren present at Christmas 2019 celebration

CHAPTER 7

HOLIDAYS, MEALS, AND SPECIAL DAYS

Christmas, of course, was a special holiday. Dad would chop down a cedar tree from the "woods," and we would decorate it with those "big bulb" lights, tinsel, and lots of ornaments. We kids remember Mom taking us to the Western Auto store in Red Bud weeks before Christmas to see all the toy selections, from which we would write our wish list. We also used the Sears catalog for our wish list. Santa always came while we were all at the St. John's Lutheran School Christmas Eve service. A treat after the service on the way out for all the participating children was a small brown lunch bag with an orange and pieces of candy. After that service, Dad would take us around town to see the beautiful Christmas decorations and lights, just prolonging our anticipation of excitement at home. Then, we'd finally get home to find and open our presents under the tree. For many, many years, we each received just one present. One Christmas, Judy had to pretend she was surprised. She knew she was getting roller skates, because one day when she went up to the highway to get the mail, there was a package from Sears & Roebuck, and it was clearly roller skates. By the way, our mailbox was on Route 159, so we had to walk the lane every day for our mail. Our address was simply RR #2, Red Bud, Illinois.

Christmas morning was no exception on a dairy farm—those cows still needed to be milked! Then it was back to church for the Christmas Day service.

Sometime around the mid 1960s, Carl and the boys fashioned a huge, lighted star for the top of the silo. It is plainly seen from highway 159 and is still enjoyed to this day as brother David still carries on that tradition.

Easter was always a special day. Mom, being the creative person that she was, even found time to make our first Easter baskets. She decorated large empty oatmeal boxes using a paste of flour and water and colorful crepe paper, even adding crepe paper ruffles. In later years we would wake on Easter morning to find a chocolate bunny waiting in our respective store-bought baskets on the dining room table. We found out later while writing this book that brother Tom would get up early Easter morning and look out the window to see where the eggs were hidden so that he had an "edge" on the rest of the clan! We hunted for real hard-cooked Easter eggs—no plastic eggs back then. With living on a dairy farm, we did not attend an Easter sunrise service, but we did go to the Easter service at the "normal" 10:30 time.

Easter 1949 with the homemade baskets:
Tom, Lois, David, Judy, Steve, John

Easter 1953: Lois, John, Judy, Tom, Roger, Kathy

Easter 1955: Tom, John, David, Judy, Kathy, Lois, Roger, Steve

On Easter Sunday in 1979, a most tragic incident occurred on the farm. Six-year-old granddaughter Kelly Guebert lost her foot in a horrible lawn mower accident. She has dealt beautifully with the situation, but not without numerous surgeries and prostheses.

Birthdays were special, but not the big celebrations that today's children have come to expect. Mom always made us a birthday cake, and there would be one present from her and Dad. Sometimes we received gifts from our godparents, but there were usually no parties. Remember, we had no grandparents to invite, and aunts and uncles' families were too large to invite everybody. Kids' parties were not even considered! Although, Judy had a 10th birthday party with some classmates, probably because she was invited to the triplet Guebert cousins' 10th birthday a few months earlier. Eunice, Eula, and Eugene were the same age in the same grade.

Kathy does remember one special Birthday/Christmas when she turned 5 years old. Dad made a small table for her. He was not noted for doing much carpentry, so for him to make a table was something special!

Kathy cutting birthday cake at table Dad made
and wearing dress Judy made

A special Christmas for the boys was the year they received an electric train set. Of course, that was one train set for all five boys! They had to share it, too, whenever they set it up, as evidenced in this photo of Kathy playing with the train! Note the old piano that Esther brought with her from her childhood home and the sewing machine that was often in use, along with the plastic curtains which were in style in the 1960s.

In this next picture the train is set up in the basement, and Steve is wiring lights for some of the little houses he made. Note the big apple butter kettle that is turned upside down underneath the table and part of the apple butter paddle that is hanging on the limestone wall above Steve's head.

Halloween was not a big deal either. Dad thought it was terrible to go trick or treating and expect people to give you candy. Plus, we lived in the country with few neighbors. However, we did still have fun dressing up in old bridesmaid dresses and assorted clothes.

One Halloween evening Judy, John, and Lois dressed up in old clothes from the trunk in the master bedroom, snuck out the back door, and knocked on the front door for a "trick or treat." They don't remember if they got any candy or if they got in trouble for getting into the trunk! Being the youngest in the family had some advantages for Kathy, as she could stay overnight with a girlfriend in Red Bud and go trick or treating with her.

New Year's Eve was a special time when we all got to stay up late. Dad always insisted you had to have pickled herring on New Year's Eve—must have been a German tradition. Uncle Honey and Aunt Esther or Vernon and Evie Muench or Rich and Twila Guebert or Mom and Dad would take turns going to each other's houses for New Year's Eve. We kids often fell asleep before the party was over.

Thanksgiving and Christmas meals usually meant turkey and gravy, dressing, mashed potatoes, corn or green beans, cranberries, jello, and pies. Mom was a great cook, so everything was good, but her dressing and pies were exceptional.

Work on the farm was very physical, so the meals were always hearty, and food disappeared fast with eight of us kids! Dad's motto was "I eat to live, not live to eat." On field work days, usually in the summer, there were four meals a day instead of the usual three. Breakfast was usually bacon and eggs for Dad and oatmeal or cold cereal for us kids and sometimes homemade Swedish tea rings or caramel rolls. The noon meal was called dinner. It seemed us kids had butter with jelly bread with almost every meal. There would be a mid-afternoon lunch of apple butter bread with summer sausage sandwiches. And Dad always needed summer sausage to go with his coffeecake. Yep, sweet and salty together is great! Supper was usually served after milking. On non-field work days, we ate before milking at 5:00 pm, which made cleaning up the kitchen much more pleasant than doing so after 8:00 at night! Common bedtime snacks were ice cream or a dish of cereal.

As you might imagine living on a dairy farm, we drank a lot of milk as we were growing up! When the gallon jug of milk was empty, we didn't have to go to the store. Just a trip to the milk house any time, day or night, to refill the jug from the big bulk tank was all that was required. The milk, of course, was not homogenized so it needed to be stirred or shook up as the cream always rose to the top. Sometimes Dad would skim off some of the cream on top of the milk pitcher for his coffee or cereal. However, we never made our own butter, ice cream, or cheese. A favorite treat was a homemade Ranger's Delight or chocolate chip cookie and an ice-cold glass of milk!

Sunday dinner was often fried chicken. That meant almost every Saturday night during the summer months, Mom and Judy butchered three chickens. Mom would cut off their heads on a chopping block near the garage, dip them in boiling water to remove the feathers, singe them to get rid of the pin feathers, clean them, and then cut them up. Early on Sunday morning Mom fried the chicken and then simmered it in the oven while we were in church. It was the best Sunday dinner. To settle the fights for the gizzards, Dad would claim them!

Growing up there was no such thing as "going out to eat," unless you count eating at church potlucks or the annual church chicken dinner and picnic! The first time we remember ever eating out in a restaurant as a family was in 1965, after Judy and Charles were married. We went to Augustine's restaurant in Belleville, Illinois, to celebrate Carl and Esther's 25th wedding anniversary and then again in 1970 for their 30th anniversary!

Carl and Esther's 25th Anniversary, 1965

Carl and Esther's 30th Anniversary, 1970

Carl and Esther's 35th Anniversary, 1975
Celebrated at 4 Flags Restaurant in Red Bud

Carl and Esther's 40th Anniversary, 1980
Celebrated at St. John Lutheran School in Red Bud

Carl and Esther's 45th, 1985, celebrated
at St. John Lutheran School in Red Bud

Carl and Esther with their grandchildren on their 45th Anniversary

The annual church/school picnic was a special day. Uncle Honey actually gave each of us a quarter to spend. There were outside games for various ages. We were able to play several games at only a nickel per game, which meant we almost always came home with four or five packs of Juicy Fruit gum! There were special treats from Dad and Mom also: Vess soda and a box of Cracker Jacks for each of us!

Even though Mom was a great cook, there were certain foods Roger and Kathy did not want to eat, like liver, brains, blood sausage, liver sausage, and assorted vegetables! So, Roger and Kathy would get sent to the "bricks" if they refused to eat what was being served! (That was the brick patio just outside the back porch.) You might be wondering how we fit ten chairs around the kitchen table. Well, we didn't! Chairs were around three sides of the table with a bench along the fourth side against the wall. A lot of kids fit on a bench!

We girls also became pretty good cooks. Judy was adept at making tapioca pudding and cream puffs, and both disappeared in a hurry. The same was true with watermelon, and boy, did it taste good on a hot summer day! (Remember, no air conditioning until the 1970s—just a box fan.) Lois and Kathy's specialty was probably pie, although we all made homemade bread and cookies, especially Ranger's Delights.

Jelly-making, peach canning, and apple-butter making were also special days. Mom made grape jelly from the grape arbor which originally hid the outhouse. She made blackberry jelly from the wild berries we picked in the woods for both jelly and pie. Dad never helped with any kind of housework, but he loved to pick those blackberries. We kids have fond memories of picking blackberries with him, even though we were bound to come home with chigger bites and scratches from thorns. It was always so hot in July when the berries were ripe. Yet, those berries were so worth it. Nothing beats blackberry pie—unless it's blackberry pie alamode!

For the peach canning, Mom would buy a bushel or two or three of peaches from Clovis Orchards near Chester, Illinois. The peaches were blanched quickly in boiling water to loosen the peelings. We girls helped Mom halve and pit the peaches. Then while Mom made the sugar water syrup, we girls packed them upside down neatly in quart jars just as Mom taught us. The jars were filled with the syrup and then sealed in a boiling water bath. We kids would eat our fill of the yummy, slimy peach peelings.

Apple-butter making was even more fun. Again, Mom would buy apples by the bushels. Our Guebert aunts would come the night before to help cut up the apples. But we kids would help by peeling the apples with the nifty hand cranked apple peeler. It was a challenge to keep the apple peeling from a single apple intact without it breaking! It's a wonder none of us got sick from eating all those peelings. The aunts and Mom would sit around a circle, kabitzing, slicing apples and tossing them into large tubs, probably the same tubs we "swam" in on the bricks when we didn't swim in the pond! The next day a fire pit was lit early in the morning, and the apple slices, water, sugar, and cinnamon sticks were put in a big copper lined kettle above the fire and cooked for hours. This had to be stirred constantly with a big, long-handled paddle stirrer for hours. Finally, when the apples had cooked down to the right consistency and color, the hot apple-butter was ladled into jars and sealed. We kids enjoyed sucking on the discarded cinnamon sticks, and afterwards, a slice of homemade bread with apple-butter and summer sausage was yummy!

Sometimes in the fall there would be pecans to gather. The trees produced more some years than others. One memorable Thanksgiving day when Aunt Norma was with us, we spent the afternoon out in the woods gathering buckets of pecans. Dad would crack the shells, and we would sit around the table picking out the nuts from the shells.

We looked forward to visiting days when we would go to visit an aunt and uncle, or they would come to see us. Sometimes we all piled in the car and drove down to Wittenberg, Missouri to visit Aunt Martha and Uncle Otto. We were not accustomed to riding much more that 15 minutes in a car, so the one-hour drive to their house over the Perry County hills meant we needed barf bags! These we made the night before from old newspapers. We thought Uncle Otto and Aunt Martha had the neatest house, up on a hill, facing the Mighty Mississippi river in the distance. They had a hand pump for the source of water in the kitchen and an old Victrola record player upstairs.

If there was any visiting going on, that meant there had to be card playing going on. Pinochle was the game, especially for the Gueberts! Dad and our uncles were all card sharks when it came to playing pinochle. You had better play well or be prepared to be called a "dumb bunny" by Dad. Hearts, rummy, euchre, and monopoly were also played, but not as much as pinochle. Sometimes we would get out the ping pong net and attach it to the dining room table. Dad was quite good at the game, and it was fun to watch him whip that ball around.

Dad playing pinochle with John, Steve, and Tom

Dad playing ping pong with Roger and Tom on the sidelines

The Monroe County Fair was always a big deal. The boys would bring their heifers to Waterloo to show at the fair and would actually sleep overnight at the fairgrounds to keep watch of the cows. They usually came home with considerable prize money. We girls entered baked goods and sewing projects and would also win prize money, but not as much as the boys! Kathy entered candles she had made one year but realized that wasn't a good idea when they melted in the heat! Kathy remembers being so excited that Dad won first place for the tallest stalk of corn one year, that she couldn't wait and ran to tell Mom who was working in the Home Extension ice cream truck. It's the little things, right?

CHAPTER 8

WORK ON THE FARM

We have made mention of farm chores and such, but decided to devote a chapter specifically to the farming operation and how it changed over the years. Dad started with horses, then a WC tractor with a hand crank, and then graduated to a WD and a WD 45, which were all Allis Chalmers tractors. Later he switched to Olivers: an Oliver 88, an Oliver 1550, and an Oliver 1650 Diesel.

The Oliver 1550

When Dad and Mom first purchased the farm in 1946, it had the following buildings: the big wooden hay barn with a lean-to attached to it which was the milking parlor, a silo about 35 feet tall, a cattle barn, the machine shed (where the community threshing machine and steam engine were stored because it was the largest shed in the area), the small feed-house where the milk cans were kept cold, a corn crib, the garage, two small chicken houses, and the house. About 1950, Dad built a large pole barn with a tin roof north of the rest of the buildings. Judy remembers when the pole barn was being built. Neighbor Quentin Klinkhardt was over helping and there was a big rain storm. He and Dad had long chats while listening to the rain hit the tin roof. The south half of the barn was for the cows, while the north half was a hay barn. That's where we kids had fun arranging hay bale houses and swinging on the rebar! The Shaeffer brothers later built an extension to this pole barn.

When the big old hay barn burned down in August of 1955, the adjacent milking parlor and the little silo also were demolished. In some ways, the fire was a blessing in disguise. A new milk house, hay barn, and silo 14 feet in diameter and 50 feet tall were built. The following year a "twin" second silo was built. In the early 1970s the twin silos did not provide enough storage. So, these were torn down. A silo 20 feet in diameter and 75 feet tall was built where the twin silos had been and was used for silage for the milk cows. A second silo was built several feet south of this one for the heifers. Plus, a larger milk-house and another cattle barn were built. Later on, grain bins and a large equipment storage shed were also built.

Twin silos and hay barn in 1968

Newer silos in 1979 with David and Alice's house on left

Silage making was a big part of running a dairy farm. The following pictures give you a glimpse of the whole process from harvesting in the field to loading it into the silo. They also show the terrific views from the top of the silo. Plus, you see how daring brother Tom was at times!

Cutting down the corn for silage

Loading the silage into the wagon

Blowing the silage into the silo

Leveling off the silage on top of the silo

View of the tractor and wagon from the top looking straight down.

View of the blower from the top

Tom walking on the eye beam between the twin silos!
He was nuts!

View of the cows and feed lot from the top

View of house from top—Notice Mom is hanging laundry
on the line behind the garage.

View of snake tree from top.
Note the stacks of straw bales by the tree.

The snake tree was called that because we think someone once saw a snake in it and because the trunk bark actually looks like snakes! The tree is still standing so you can check it out for yourself! It has to be over 100 years old. In fact, it may be the oldest Osage orange tree in southern Illinois.

Carl only had about 16 dairy cows when he first started farming, and the milking was done by hand into a bucket. He never had any bulls, using artificial insemination from the very beginning back in 1939. The milk was put into milk cans and set in several inches of water to be kept cool until it was time to haul them up to the end of the road to be picked up. This all changed in 1955 when a new milk-house was built with a state-of-the-art stainless steel tank to house the milk and keep it cold. A big tanker truck came every other day to take the milk to the Pevely Dairy to be processed. The following newspaper article reflects that change. When brother David retired in 2008, he had up to 50 head of dairy cows.

Cow Herd, August 1968

THE WATERL

ON FARM OF PRESIDENT OF RED BUD LOCAL

UPPER LEFT — Carl E. Guebert, president of the Red Bud Local of the Sanitary Milk Producers, pictured in his milk barn with some of his Holstein cows.

UPPER RIGHT — In this photo, Mr. Guebert is pictured as he puts milk from the milking pail into the vacuum cooling tank in his milk house.

BELOW — Mr. Guebert does not sell all of the milk produced by his herd of dairy cows, as attested by this picture of his children, seated around the kitchen table, each with a glass of wholesome, health-giving milk. Around the table from the left, are Judy, Steven, John, Tommy, David, Roger and Lois. The baby of the family, Kathie, is camera shy, and refused to pose on the picture.

—Times Staff Photos

Milk can days are over on the farm of Carl E. Guebert, who resides about two miles south of Hecker. Since November 1 of the past year, Mr. Guebert has a vacuum cooling tank in his milk house. The milk is taken from his milking pails by vacuum into the tank, and from the tank it is pumped into the William Braun tank truck for delivery to the dairy.

Mr. Guebert has a herd of 16 Holstein cows and his records for last year showed that he produced an average of 11,020 lbs. of milk per cow, with an average of 427 lbs. butterfat. He farms a total of 150 acres, on which he produces wheat, corn, soybeans and alfalfa. In addition to his cows, he maintains a flock of 400 hens.

In addition to his duties on the farm, Mr. Guebert finds time to serve his church and his school district. Besides being president of the Red Bud local of the Sanitary Milk Producers, Mr. Guebert is a member of the Board of Education of Waterloo Community Unit School District No. 5, and also serves on the Board of Education of St. John's Lutheran School at Red Bud. Mr. Guebert and his family are active members of St. John's Lutheran Church at Red Bud.

AUG • 68 •

Dad at work

Twin silos and milk tanker truck in front of the milk house

There were very strict rules concerning the milking operation, and our farm had a very high rating. A certified milk tester (Frank Schultze) would actually come to the farm and stay overnight so that he could test samples from both the evening and morning milking. We remember watching Mr. Schultze use his centrifuge machine to spin the testing bottles to check for butterfat content while he listened to classical music as he worked.

Up until the 1960s the cows always grazed in the pasture. So, someone, usually one of us kids, would go down the lane past the snake tree and call "Soooie" to get the cows home for milking. Interestingly, the cows would walk home single file, creating a worn path that we pretended was an Indian trail! Later on, Dad learned that keeping the cows close to the barn and feeding them more silage and hay instead of letting them graze was more productive. This eased the workload and led to higher milk production. Now, no one had to call the cows home, keep the fences mended, or maintain the water trough out in the pasture. Plus, that eliminated the problem in the springtime of the milk smelling and tasting like onions from the wild onions the cows ate! The cows still had an area close to the barn where they could walk around to get some exercise.

Obviously, the milking was done on a daily basis, twice a day, usually about 6 am and 5:30 pm. There were early morning chores before school and in the evenings, getting ready for milking. The cows needed to be herded into the holding pen for milking. Another chore was feeding the remaining "dry cows" and heifers hay bales and silage and also feeding the calves. In the evenings, straw would be spread for bedding. While one person was starting the milking, this was a good time to clean the lot since all the cows were in the holding pen ready to be milked. So, another person would scrape the lot with a tractor and bucket and load it into the manure spreader. This was later spread out on the fields for fertilizer. Later on, a manure pit was dug, and the liquid manure was spread on the fields with a slurry or honey wagon.

Dad scraping the cow lot and loading it into the spreader

A load of manure ready to be spread on the field

The barn that had burned down provided many memories:

- Loading bales into the hay loft, first with a pulley system and a huge bale hook and then later with an auger.
- Lois and Judy daily sweeping the concrete floor of the milking area of the barn and sometimes roller-skating on the clean rough concrete.
- Lots of pigeons roosting in the barn's rafters!
- Kittens hiding in the hayloft.

The main crops that were grown on the 150-acre farm were corn, wheat, soybeans, and alfalfa. The whole stalks of corn were cut and chopped up while still green and put in the silos for silage feed for the cows. The wheat was combined and sold while the stalks were baled for straw for bedding for the cows. The soybeans were combined and sold. The alfalfa was cut and baled about four times during the summer, roughly every 30 days, and used for hay for the cows. In early years, Dad planted sorghum cane which we called sugar cane. It was hauled into Red Bud to be made into molasses.

Hay baling took many hands and was labor intensive. We think Dad was pleased to have five sons! However, he put Judy to work driving the tractor until the boys were old enough to do so. Then it was usually Roger driving the tractor, John or Steve stacking, and David and Tom throwing the 40-pound bales onto the wagon.

Dad loading hay, 1968
Where are the boys? Here they are! . . .

Steve throwing bales onto hay wagon, 1968

David throwing hay bale to Tom, 1968

Tom Stacking Bales on Wagon, 1968

Tom and Steve on the hay wagon, 1967

The wood cutting was labor intensive, also, and dangerous as evidenced by the accidents previously mentioned. First, the trees had to be cut down. Then the wood was cut into logs with a tractor and buzz saw. The Osage orange (also called hedge apple) trees were cut down for firewood because they burned long into the night and for fence posts because the wood did not rot easily. The hedge apples were even useful as the hogs liked to eat them, plus, a few hedge apples thrown into the crawl space helped to keep the bugs down. However, the Osage orange wood was really hard to cut, and that is the wood Dad was trying to cut when the chainsaw jumped out at him and cut a deep gash into his knee. He was by himself when this happened, which made the situation even more dangerous! Somehow, he managed to get back up to the house. Mom came home from picking us kids up from school and found him sitting in his chair with a bloody knee! She promptly took him to the doctor; thankfully the injury only required stitches, but he was laid up for a while.

There was always a vegetable garden. Dad had a neat system for planting the garden. He would lay down a wooden board that he would walk on and then dig a trench next to it for planting the seeds in nice straight rows. The main vegetables were green beans, tomatoes, cucumbers, asparagus, lettuce, sweet corn, and potatoes. Dad planted three long rows of potatoes behind the granary barn (machine shed) along the fence leading to the snake tree. Judy recalls planting and harvesting buckets and buckets of potatoes. We stored them in a large bin in the basement fruit cellar. Brother Tom remembers Dad picking a tomato and biting into it straight from the vine. Some of us remember doing the same. They were so good. The tomatoes that made it into the house were served sliced with a sprinkling of salt, not sugar! We also grew rhubarb, and of course, grapes.

Obviously, the main livestock was the cattle, and the cows all had names. Frieda, the best producing cow for a while, was Judy's favorite. We also had chickens and hogs until the early 1960s. The hog sheds actually sat on skids so that they could easily be moved from one location to another. One side note about the hogs is that besides the normal feed that they were fed, they were also fed vegetable peelings, food scraps—SLOP!, and cookies!

A few times Dad got cookies from Schwartz's Bakery that were stale and broken—culls that could not be sold. The men that delivered the cookies were black—the first black people we encountered. Remember, we led a sheltered life! Dad stored the cookies in the attic of the machine shed. Before the hogs got the cookies, though, we kids climbed that ladder and searched through the gunny sack bags for any unopened packages for ourselves! There were even marshmallow chocolate cookies that we never had before. They didn't taste stale to us!

Butchering days were big days. In the fall, once the weather turned cold, usually in November, it was time to kill a hog or two. Uncle Honey, Uncle Bank, and Eugene Bruins usually came out to help. First, the hog was shot, and the blood was drained into a big bucket. This, along with corn meal and seasonings, was cooked and stirred to be made into blood sausage. In Red Bud, it was called "webconbrot," also now called klump. The liquid mixture was then poured into cloth bags which Mom had sewn. It solidified as it cooled. Mom would fix it for breakfast; diced and fried, and served with liver sausage mixed with oatmeal, like grits. Some of us kids ate it, but some of us chose not to!

The men set up a big wooden table in the machine shed where the meat was cut up. The hams were salted and hung up in the smokehouse shed (the little feed house) to be cured. Much of the meat was ground up with a big hand-cranked meat grinder and stuffed in the cleaned out intestine casings for pork sausage. Using the intestine casings sounds rather gross now, but it was an efficient use of resources, and none of us ever got sick from the food we ate! While the menfolk were busy with the butchering tasks, the womenfolk were busy preparing a huge spread of a meal for a hard, long day's work.

After butchering, Mom would always can links of pork sausage. She fried it down in the oven and then placed links in quart jars, poured the grease over them, sealed them, and turned the jars upside down. We stored them in the fruit cellar in the basement. It was a fast and delicious meal she would fix for company many a time.

In the early years before combines, a threshing machine was brought in to harvest the wheat. This was very labor intensive, so extra hands were needed, and a big meal was prepared by the womenfolk similar to what was done for butchering. First, the wheat was cut down and tied into bundles, either with a horse-drawn or tractor-drawn binder. The bundles were stacked into shucks, appearing to be mini teepees in the field.

After the wheat was sufficiently dry, the bundles were pitched off by hand one bundle at a time into the steam powered threshing machine which separated the wheat kernels from the straw stalks.

David and Alice took over the farm when Dad retired in 1979. The 1980s were tough for them as they had $2,500 in farm payments every month! This meant they needed higher production from the cows. Through hard work, being diligent in cleanliness, purchasing new milking equipment, and other means, in a few years they achieved their goal. The herd was on top in Monroe County for over 20 years and was in the "Top 10" of Illinois's over 3,000 dairy herds.

Aerial view of farm in the 1970s

Aerial view of farm in the 1990s

CHAPTER 9

SHENANIGANS AND FAVORITE SAYINGS

Kids will be kids and even the quiet Guebert kids were no exception. We did not always behave as we should. That's what we mean by "shenanigans!"

There was the time Judy and John argued at the top of the stairs, and Judy came tumbling down! Fortunately, she wasn't seriously hurt, but did need stitches. Then there was the time John and David were playing cowboys and outlaws. The chair slipped as John was setting David up for a hanging, and there almost was an actual hanging!

There was also the time Roger and Kathy got caught by John smoking behind the barn. Kathy said it was all Roger's idea, that he was good at getting her in trouble! Surprisingly, Dad did not get real upset about it. However, Roger later experimented with some matches and made burnt designs on the master bedroom window ledge but did not admit doing it. That is when Dad hit the roof because Roger had lied to him. Dad sent all the rest of us kids down to the pond to swim and took a leather belt to Roger's behind. Roger had a knack for getting into more trouble than the rest of us. He once got so mad that he threw a boot through a porch window! Mom always called him the "little dickens." He was always so cute, and we all loved him just the same.

Roger was not the only one to get in trouble for throwing something. John threw a brick through the porch and into the dining room window from the brick patio! He said it was the only spanking he ever got.

Dad was a cigarette smoker and occasionally chewed tobacco all the while we were growing up. In his defense when he started smoking it was not linked to cancer as it is now. Years later when young granddaughter Jenny told Grandpa that he should not be smoking, he quit! This also happened to coincide with the "Stop Smoking Day" on KMOX radio.

Being the youngest with five older brothers, Kathy was often subjected to their teasing. She remembers crying to Dad that they called her fat, and he replied that she was pleasingly plump, not fat. For some reason that didn't make her feel any better!

While Mom was helping Dad with the milking after supper, Judy and Lois spent much of that time arguing over who would wash and who would dry the dishes! Once in a while, they hadn't even started the dishes when Mom got back in from milking. With ten in the family, there were lots of dirty dishes and no automatic dishwasher.

Some things are funny when you look back at them but not when they were happening. Such was the case when Roger was hauling a load of shelled corn along Highway 159. He evidently got too close to the shoulder of the road, and all the corn tipped out! As you might imagine, Dad was not real pleased with the situation!

Evidently the silo was a big temptation for some of us kids. A couple of us were brave enough to climb the original 50 feet silo on the outside ladder that had only open bands around you. Mom wasn't too pleased to see 8-year-old Judy up there waving to her at the back door one day. Another time, Dave and Roger, about 10 and 8 years old, thought it would be cool and fun to pick some little green apples and go up into the silo to eat them!!! Maybe they thought they'd get caught eating little green apples.

These many years later, we find out the two youngest boys, Tom and Roger, were real dare devils. They both actually climbed all the way to the top on the back side of the silo by climbing the rods that hold the silo together. They did this when the silo was almost completely filled. On silo filling days, we kids had the job of packing down the silage as it was being blown up and into the silo. We'd be barefoot and walk in circles, stomping it down. The silage smelled just like green olives—no wonder some of us don't like olives. Back to the story—Roger tells us when the silo was almost filled, to about waist high, the boys at the top were throwing down corn chunks at the boys below. So, Roger's game plan was to surprise them by climbing the back side and hop over the top! Bet Mom and Dad never knew this! No doubt the angels were with the boys making sure their fingers and toes had good grips.

Dad had some favorite sayings. If we girls complained about not knowing what to wear, he always said, "Wear the pink one," whether we had a pink dress or not. If we kiddos got a little noisy, he would simply say "Pipe down" or "That's enough," and we stopped immediately. Sometimes in the middle of evening milkings, Dad would simply say "I got to go," and he'd leave to attend a meeting, leaving the boys to finish up milking. At meetings when issues were being discussed, he was known to say, "Now wait just a minute."

Dad would say that you can fix about anything with bailing wire and/or duct tape, though he used bailing wire more, particularly for fences and some hand tools. Usually, once he made such repairs, he would say that should hold until it needs fixing again! Another one of his "tricks" was to apply oil to his work shoes. He also had buckets of old motor oil that he applied to hoes, shovels, plow shears, disks, etc. after cleaning off the dirt to keep them from rusting. Long before weed eaters and Roundup came along, Dad used long handled scythes for clearing fence rows.

If you got up really early in the morning, you would often find him eating a bowl of cereal, while listening to KMOX, and reading either the Bible or Reader's Digest before he went out to do the milking. He loved to listen to Hawks basketball, and both he and Mom loved to listen to Cardinals baseball with announcer Harry Carey.

Dad did not understand why there was a need to shop anywhere else but in Red Bud. There were two typical old-time general stores (Schrieber's and Voss's) where you could find everything from patterns and fabric to jeans and shoes to groceries and cleaning supplies to furniture and hardware. There was also Rathert's clothing store and Dehne's grocery store. Dehne's was located on the north end of town and was a regular stop on the way home from school. You could charge everything until the end of the month and then write one check for all of it. (Credit cards did not yet exist.) Another shopping option was local auctions. Mom enjoyed going to these and sometimes bought furniture or dishes at them. We enjoyed going to them also.

We rarely got sick, but when we did, there were two general practice doctors right in Red Bud and a hospital. They didn't take appointments back then. Instead, you sat in the waiting room (for hours and hours it seemed) until it was your turn to see the doctor. Some of the visits to the doctor were actually for bad cases of poison ivy or poison sumac. That was not fun! We did not have health insurance. We never even heard of it back then, and medical costs were much lower.

Dad was nearsighted, and evidently all eight of us kids inherited nearsightedness from him as we all ended up wearing glasses! Dad said that he probably should have had glasses a lot earlier than he did, as once he got them, he realized that Christmas tree lights were not supposed to have that fuzzy glow around them, and ties had designs on them! Kathy remembers having that same experience before she got her glasses in fourth grade! Once we started wearing glasses, we had an easy test to know if we needed stronger lenses. If we couldn't read the hymn numbers on the board in church, we knew it was time to go back to Dr. Matthews in town for new glasses!

Both Mom and Dad were raised in strict German Lutheran families. So that is how they raised their children. They did not show affection. Growing up, we don't remember getting hugs or kisses. In fact, Judy remembers it feeling strange for Mom to actually hold her hand crossing the street one time in downtown St. Louis. Even so, we knew we were loved. Dad often referred to Mom as his sweet bride.

CHAPTER 10

Memories from the Grandchildren

Even though this book is mainly about us growing up on the farm, we thought it would be good to include a few of the memories from Esther and Carl's grandchildren.

Jodi's memories:

- Climbing the ladder to the barn attic and then jumping into the corn/grain bin and playing in it. Wonder if Grandpa knew about this!
- Playing in the hay bales and looking for kittens in the hay barn.
- On hay bailing days—watching the bales go up the conveyor belt to the barn.
- Baby calves sucking our fingers with their rough tongues.
- Going out in the woods with Grandpa to the pecan tree and picking up buckets of pecans in the fall of the year.

Randy remembers:

- On holidays—"Uncle Johnny's moto bike." Later Uncle Dave and Roger also had motor bikes. They'd give us rides up the lane and back, and out to the woods.
- Pumping water from the cistern.

- Seeing newborn calves.

Jeremy's memories:

- My kindergarten class, under Mrs. Moellering, took a field trip to Grandpa's farm. I was so proud and couldn't wait to take my friends out to the hay barn. But we had to wait till the tour was over.

1986—Immanuel Lutheran School Olivette Kindergarten class

- I also brought several friends to the farm for a sleep-over once, and we played ball on the big open front yard.

Mark remembers:

- Grandpa letting me drive the tractors while sitting in his lap.

- Working the hay fields in the summer. It was hard work, but I enjoyed spending time with Grandpa in the fields and dinner in the evenings.
- I also remember some mischief Randy and I got into one time throwing snowballs at the cows.
- Taking my kids Timmy and Elisabeth to see the calves and letting them suck the kids' fingers and how much the kids smiled.

Jenny's memories:

- Holidays at Grandpa and Grandma's.
- Playing on the stairs, sliding down the banister.
- Eating Grandma's pies (mmm).
- Couldn't wait to go play in the hay bales.
- Letting the baby calves suck my fingers.
- Playing with the kitties from under the porch or by the milk house.
- When I was really young, Grandpa letting me crawl on his lap when in the truck or tractor to "drive" and going with Grandma to deliver food to Grandpa when he was planting or harvesting.
- The men sitting around the dining room table playing pinochle at family gatherings.

Becky remembers:

- Grandpa and Grandma had a HUGE garden on the southeast area of their yard; they had green beans, potatoes, rhubarb, strawberries... there were peach trees along the garden, and Grandma also had a raspberry bush and asparagus patch. There was also an apple tree in another part of the yard, and I remember Grandpa and my dad planting a HUGE patch of sweet corn; Grandpa would fill a cart with corn and take the husks off on the back porch, and I helped my mom and Grandma in Grandma's kitchen, cooking the corn and putting it in freezer bags. I remember getting blisters on my hand from using a paring knife to cut kernels off the cobs in a metal pan.

- Grandma always kept grape bubblegum as a treat for us in the top left drawer of the dining room buffet. On the rare occasion she was out of gum, the drawer still smelled like grape. And speaking of grapes, I also remember Grandma making grape jelly from her grape vines in the yard; the best treat was when she'd keep some sweetened grape juice for us to drink. She also made blackberry jelly; I remember riding the 5-wheeler down the dirt road to the pond and picking wild blackberries with Grandpa.
- We also loved going to the pond closer to Grandpa and Grandma's house and feeding the catfish. Grandpa fed them dog food; we loved watching the fish gulp the food with their huge whiskered mouths.
- Spending the night at Grandpa and Grandma Guebert's house was a rare treat for us; I loved the idea of getting to sleep upstairs since I hadn't been to many houses with an upstairs, but it was also scary because you could hear mice or birds in the ceiling! Now my boys sleep up there and thankfully they haven't ever said it was scary

except once in a while during storms! I was also torn about the cellar... I thought it was really cool how you could get in and out of the basement through the really old cellar staircase, and I loved the stone walls and the pantry room where Grandma had rows and rows of canned goods. But the cellar was/is also very scary, especially the back room where you have to go to the middle of the room in the dark to pull the string on the light—to this day I won't go down there at night by myself, and really I'm not a huge fan of going down there in the daylight!

- Grandpa and Grandma's house seemed enormous to a little kid. It felt like we fit 50 people in there at Christmas; I remember running laps through the house with my cousins, and eating at the "kids' table" in the kitchen where I think we managed to fit a dozen kids around a table I now know only fits 4–6 adults comfortably! I loved the huge dining room table, and I remember Grandpa hosting some vicious pinochle games there.

Karla's memories:

- I remember Grandpa feeding dog food to the fish in the pond; he wouldn't fish, just feed them.
- We would go over to Grandma and Grandpa's house while Mom and Dad were milking. I would lay on the floor watching TV, and Grandpa would stand over me and shake me with his legs and make me laugh.

- Grandma was always the one we went to when we got hurt (she always used a glass vial for our wounds, but I never knew what it was; apparently it's mercurochrome).
- Grandma was always there to pick me up from sports games or practices and got us ready for school in the morning.
- Grandma always knew when Traci was faking sick so she wouldn't have to go to school!

Jackie's memories:

- I remember being so excited to be invited to the big boy table to play euchre during a Christmas gathering. I never thought Grandpa would let me in and possibly mess up the game!
- Whenever I picture Grandpa, it is in blue jean overalls. I can't imagine him in anything else.

Annie remembers:

- I have so many fond memories of visiting the farm. Since we never lived close, we always got to spend the night! My brother and I would sleep in the twin beds in the upstairs west bedroom. The sound of birds would wake us in the morning!
- The stairway was used as jail when we cousins would play cops and robbers.
- We would also play house in the hay in the barn.
- I always enjoyed watching the milking process.
- Grandma would put a safety pin in the middle of the top sheet to aid in making beds! After Grandpa passed away, Grandma, being the practical person she was,

would alternate sides of the bed, so that she used both sides of the sheets and so the bed was worn out evenly!

The following article sums up ours and our children's memories of going back to the farm. It was written by Judy and appeared in the February/March 2002 issue of Country magazine.

Back to the Farm

MY HUSBAND, Charles, and I had the opportunity to watch our granddaughters Jordan and Jenna, ages 4 and 2, for 5 days while their parents went on an anniversary getaway.

I grew up on a farm about an hour's drive from where we presently live, and I figured these city girls would enjoy spending one night there. My brother David now runs the family dairy farm, and my mom still lives in the farmhouse where my seven siblings and I were raised. While Mom's activities have recently been limited, she welcomed our overnight visit and the chance for more bonding time with her great-granddaughters.

We know the girls loved it, but more than that, I'm still gloating over the fact we provided an experience they'll probably never forget. How many kids nowadays get to go to a real farm, and how many can say they slept in the same room where their grandma slept as a child?

They saw a mother cat caring for her kittens. Jordan and Jenna loved carrying the kittens around—and watching Grandpa rescue one of them from a tree.

Great-Grandma drew sketches of animals for them on scratch paper. I only wished I would have saved those sketches, but Mom says she'll draw some more and mail them to the girls. Our daughter Jodi will love the sketches as much as the girls.

When it was time for bed, we said our prayers together, and I tucked them in. A few minutes later, I heard little footsteps come across the creaking hallway. Jordan came to the side of my bed and whispered in my ear, "Grandma, I just had to tell you what a really fun day this was." She then went back to her bed and fell right asleep.

The next day was filled with picking blackberries, watching the bulk tank truck empty Uncle David's milk tank and petting the calves.

The girls didn't want to go home, and to tell the truth, neither did I.

My first day back at work, all I could hear was that little whisper, "Grandma, I just had to tell you what a really fun day this was." I was so touched.

I mailed a letter about the visit to one of my sisters and sent Mom a copy. Evidently, she cherished our visit as much as we did, because she told me she read it over at least eight times.

—Judy Meyr
St. Louis, Missouri

BERRY BONANZA. A visit to their great-grandma's farm meant picking blackberries and other fun activities for Jenna and Jordan.

CHAPTER 11

Family Tree

The Family of Esther Anne (Werner) and Carl Edward Guebert

(4/6/1918 – 5/31/2007) (1/4/1914 – 2/23/1990)

Note these family trees are current as of December 2019. An asterisk after a name denotes that the family member inherited the polydactyl gene.

Judith Catherine Guebert* (5/1/1941) and Charles Leon Meyr (10/10/1939), married 5/11/1963. Judy is retired from Edward Jones Investment Co. and keeps busy with volunteer work at Rankin Jordan Pediatric Hospital and at church. Charles is retired from Hummert Seed Company, is also active with church, and plays pinochle whenever he can! They reside in St. Louis County, MO.

Jodi Lynn Meyr* (8/14/1964) and John Henry Hose (1/5/1965), married 7/21/1990, reside in Ballwin, MO 63021.

Jordan Marlene Hose* (3/5/1996)

Jenna Catherine Hose* (4/28/1998)

Jessie Lynn Hose (9/10/2001)

Jolene Frances Hose (9/13/2003)

Randall Charles Meyr* (6/28/1966) and Christine Ann Roth (3/28/1970), married 7/13/1996, reside in St. Louis County, MO 63114.

Jacob Charles Meyr (2/6/2000)

Carrie Ann Meyr* (2/28/2004)

Jeremy James Meyr (5/14/1981) lives in High Ridge, MO 63049. His special friend is Robin Perry.

John Paul Guebert (7/19/1942) and Judith Fish Stuckstede (7/27/1947), married 9/3/1971. John retired from NCR Corp. after 41 years, and Judy kept busy as a full-time homemaker and mother of six children! They spend much of their time now traveling and enjoying their grandchildren. Their home is in Arnold, MO.

Mark Edward Guebert (4/8/1969) and Laura Marie Meschke (3/28/1973), married 5/19/1994, divorced in 2011. Married Vicki Lynn Strauss Russo (7/16/1969) 9/22/2012. They live in House Springs, MO.

Timothy Edward Guebert (8/21/1997 – 4/5/2003)

Elisabeth Marie Guebert (11/26/2000)

Alexander Scott Guebert (4/22/2004 – 12/6/2006)

Stepchildren: Dominic Joseph (D.J.) Russo (4/15/1992)

Greg Russo (11/25/1994)

Miranda Russo (2/13/2000)

Kimberly Ann Guebert (12/21/1972) and Michael Patrick McMullins (5/31/1963), married 5/1/1999. They reside in Arnold, MO.

Jonathan Cody Guebert (9/25/1989)

Kelly Renee Guebert (12/21/1972) and Richard Dean Grosbauer (11/27/1967–10/27/2012, cancer), married 9/12/2009. Kelly's home is in Arnold, MO.

Gregory Scott Guebert (11/15/1975) lives in Arnold, MO.

Christopher Paul Guebert (5/18/1979) and Clare Louise Krueger (4/11/1980), married 6/6/2009. They reside in St. Louis, MO.

Conrad Matthew Guebert (3/5/2011)

Cable Christopher Guebert (10/23/2013)

Corbin Christopher Guebert (10/23/2013)

Jaime Linn Guebert (7/6/1982) and David Matthew Cooper (4/14/1982), married 8/10/2001. Their home is in Chula Vista, CA.

Reese Jon Cooper (9/20/2005)

Brady James Cooper (6/8/2007)

Finn Gregory Cooper (2/21/2012)

Evelyn Grace Cooper (5/12/2017)

Lois Rose Guebert (2/10/1944 - 3/22/2010, cancer) and Joel Alan McQueen (10/25/1940), married 6/10/1973. Lois was a home economics teacher and then a kitchen designer for 10 years. Joel retired from Monmouth College and resides in the home he and Lois built in Galesburg, IL. We sometimes joked that Lois was like a kitten with several lives. She survived kidney cancer when she was only 5 years old. In 1981, in a tractor accident, she was crushed by the back wheel of the tractor and miraculously made it through that. Then in 2005 she was diagnosed with bone cancer and had a section of her right femur bone removed. She went through a period of remission during which we three sisters and our husbands took a memorable trip to Alaska. She succumbed to the disease in 2010, shortly after her 66th birthday. We feel that the Lord called her to heaven much too soon, but we do feel blessed to have had her on this earth as long as we did.

Steven Raymond Guebert* (7/21/1945) and Carol Eleanor Walter Jenson (12/4/1944), married 4/7/1989. Steve retired from working for the Farm Credit Administration in McLean, Virginia, and now enjoys gardening and doing some weather reporting. Carol did secretarial work and keeps busy with family activities and craft projects, including making banners for their church. They live in Brenham, TX.

Stepchildren:

David Christopher Jenson (9/26/1974). David also lives in Brenham, TX.

Phillip Alan Jenson (9/27/1981) and Kristin Jane Gordon, married 10/24/2009. Their home is in Herndon, VA.

Ella Jenson (4/28/2016)

David Allen Guebert (10/5/1946) and Alice May Driver (1/3/1952), married 8/24/1974. Since retiring from the dairy farm, David and Alice take a very active role in raising their grandchildren and also have done quite a bit of traveling. They still live in the house they built on the farm and rent out the farmland.

Becky Jean Guebert (7/18/1978) and Kevin Andrew Biffar (8/8/1977), married 10/16/2004. They live in Carl and Esther's farmhouse, and we are happy to see the home remain in the family.

Jakob David Guebert (3/2/2007)

Josef Carl Guebert (4/20/2010)

Traci Anna Guebert (12/30/1980) and Alan Raymond Proctor (1/11/1979), married 10/5/2002. Traci and Alan live in the country near Ames, IL.

Logan Allen Proctor (1/10/2005 – 1/12/2005)

Madison Alice Proctor (1/10/2005)

Hailey Cecelia May Proctor (10/21/2006)

Kaydence Elinor Proctor (6/5/2012)

Karla Lynn Guebert (9/30/1984) and Gabriel Gregory Hernandez (5/28/1986), married 9/17/2016. Karla and Gabe live in Lebanon, MO.

Noah David Hernandez (1/22/2019)

Thomas Carl Guebert* (9/19/1947) and Barbara Edith Osterhage (7/23/1948), married 8/16/1969. Tom retired from Country Companies Insurance Co., and Barbara retired as a legal secretary. They do extensive traveling and are active in the church and in the community, particularly with Rotary International. They live in Carlyle, IL.

Jennifer Christine Guebert* (1/25/1972) and Michael Doane, Jr. (5/13/1973), married 7/6/2014. Their home is in Manchester, MA.

Stepchildren: Finnegan Porter Doane (3/13/2004)
Ava Marin Doane (1/31/2006)

Jeffrey Thomas Guebert (11/24/1975). He lives in Norwalk, CT.

Roger Dale Guebert (10/30/1948) and Joyce Marie Posey (11/17/1951), married 9/1/1972, divorced in 2000. Roger retired from Golden Dipt Breading Co. He survived both the Vietnam War and cancer of the jaw and is active with the Veteran's Association and enjoys gardening. He lives in Millstadt, IL, and his special friend is JoAnn Suemnicht.

Jaclyn Marie Guebert (2/14/1977) and Scott Anthony Maxwell (7/25/1970), married 7/26/2003. Their home is in Smithton, IL.

Addison Lee Maxwell (7/18/2006)

Jackson Scott Maxwell (4/15/2010)

Scott Edward Guebert (4/28/1981) lives in Millstadt with his dad.

Kathryn Anne Guebert (12/24/1951) and Thomas James Murray (4/26/1950), married 8/18/1973. Kathy retired from teaching home economics, and Tom retired from Concordia Publishing House. Square dancing has sort of become their second career as Tom is a square dance caller! They live in St. Ann, MO 63074.

James Carl Murray (5/12/1978) and Jennifer Marie Hodits (1/4/1977), married 6/4/2005. They reside in Fenton, MO.

Connor James Murray (8/25/2007)

Colby Joseph Murray (4/23/2010)

Carter William Murray (5/5/2012)

Anne Marie Murray (5/13/1981) and Grant Erwin Hedeman (7/27/1980), married 8/14/2004. They reside on a farm outside of Lockwood, MO.

Molly Mae Hedeman (4/12/2010)

Ardrew Terry Hedeman (4/28/2013)